Explicit Symmetry Breaking in Electrodynamic Systems and Electromagnetic Radiation (Second Edition)

Explicit Symmetry Breaking in Electrodynamic Systems and Electromagnetic Radiation (Second Edition)

Dhiraj Sinha
Plaksha University, Punjab, India

Gehan A J Amaratunga
University of Cambridge, UK

IOP Publishing, Bristol, UK

ISBN 978-0-7503-5132-4 (ebook)
ISBN 978-0-7503-5130-0 (print)
ISBN 978-0-7503-5131-7 (myPrint)
ISBN 978-0-7503-5133-1 (mobi)

DOI 10.1088/978-0-7503-5132-4

Version: 20220901

IOP ebooks

British Library Cataloguing-in-Publication Data: A catalogue record for this book is available from the British Library.

Published by IOP Publishing, wholly owned by The Institute of Physics, London

IOP Publishing, Temple Circus, Temple Way, Bristol, BS1 6HG, UK

US Office: IOP Publishing, Inc., 190 North Independence Mall West, Suite 601, Philadelphia, PA 19106, USA

To Jeffrey Goldstone, the pioneer of spontaneous symmetry breaking.

The Centipede

A centipede was happy quite, until a frog in fun said 'Pray, which leg comes after which?' This raised her mind to such a pitch she lay distracted in the ditch considering how to run.

—Anon

Contents

Author biographies

Dhiraj Sinha

Dhiraj Sinha studied electrical engineering at the Institute of Engineering and Technology, University of Lucknow, India and Telecommunications at ENST de Bretagne, Brest, France. He earned a PhD in electrical engineering from Cambridge University, UK. During the latter part of his stay at Cambridge and as the CTO of Smantenna Ltd, a thin film based antenna startup, he worked on research and commercialization of thin film based antennas. This was followed by a brief stint at Oscion, a technology consulting firm, following which, he joined the Singapore University of Technology and Design, Singapore as a Research Fellow. Subsequently, he worked as a Postdoctoral Associate at Massachusetts Institute of Technology before joining Plaksha University as a faculty member. His research interests are converged towards radiating structures with low degrees of freedom. In that context, he has explored ultra-small antenna design, sensors and actuators, besides investigating thermodynamics and electromagnetic field effects in microscopic systems with low dimensions.

Gehan A J Amaratunga

Professor Gehan A J Amaratunga FREng., FIET, CEng., obtained his BSc (1979) from Cardiff University and PhD (1983) from Cambridge, both in electrical engineering. He has held the 1966 Professorship in Engineering at the University of Cambridge since 1998. He currently heads the Electronics, Power and Energy Conversion Group, one of four major research groups within the Electrical Engineering Division of the Cambridge Engineering Faculty. Professor Amaratunga's research is in the broad area of materials and technologies for electrical energy and power. It intersects electrical and electronic engineering with chemistry, physics, materials science and information systems. He has an active research programme on the synthesis and electronic applications of carbon nanotubes and other nanoscale materials. His current research is focused on integrated power conversion circuits.

Professor Amaratunga was elected a Fellow of the Royal Academy of Engineering in 2004. In 2007 he was awarded the Royal Academy of Engineering Silver Medal 'for outstanding personal contributions to British engineering'. He has published over 500 archived journal and conference papers (15000 citations) and is an inventor on 32 patents.

IOP Publishing

Explicit Symmetry Breaking in Electrodynamic Systems and Electromagnetic Radiation (Second Edition)

Dhiraj Sinha and Gehan A J Amaratunga

Chapter 1

Introduction

Symmetries are integral aspects of nature and natural phenomena. The Sun and Moon appear disc shaped to the naked eye and closely resemble a sphere in physical shape. The horizon appears round, representing the outline of the Earth's spherical silhouette. The heavenly objects follow a symmetrical path—there is day after night and night after day; the seasons follow a periodic pattern, where the rains follow prolonged dryness and are then followed by dreary winter and spring; and low tides follow high tides as time flows. Symmetry is inherently expressed in all living beings with variegated forms and nuances. The corolla of a flower follows a geometric symmetry (figure 1.1). Symmetry is also present in patterns on snowflakes (figure 1.2), where spherical droplets are condensed together in a certain shape, often resulting in sixfold radial symmetry. Symmetries are also observed in sand dune shapes, shell patterns, bird feathers and organ shapes, which are all beautiful examples of symmetries. Symmetry in crystals is well-known and primarily determines the structure of the majority of solids. For example, quasi-crystals are special kinds of crystals made of symmetric patterns that do not get repeated (figure 1.3).

Symmetry is seamlessly intertwined with the laws of nature and almost all natural phenomena can be correlated to some kind of inherent symmetries. The pattern of potential energy in gravitation or electromagnetism has some kind of symmetry. For example, the contours of potential around a charge have rotational symmetry (figure 1.4).

Despite the reflections of symmetry manifested in multiple forms and dimensions in nature, symmetry and symmetry breaking are seamlessly intertwined. The beautiful patterns of petals in flowers disappear, fading away spontaneously with time, along with their sweet fragrance. Ordered states spontaneously break down into states of disorder and entropy increases; living beings are born, grow old and eventually die and disappear, never to be seen again. Snowflakes melt and disappear. Time flows in

Figure 1.1. The symmetric pattern of a dahlia corolla. Credit: (Wolfram Demonstrations: http://demonstrations.wolfram.com/GrowDahlias/)

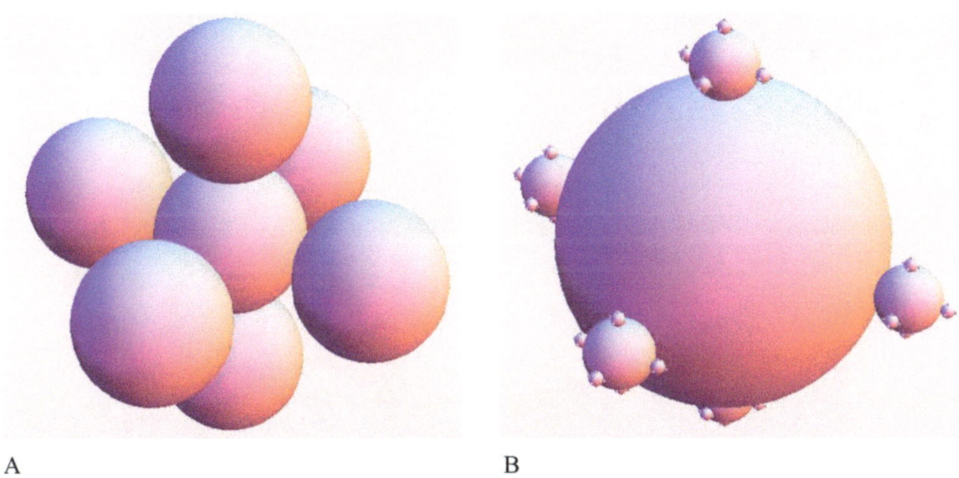

Figure 1.2. Spherical droplets, as shown in (A), can condense and begin to form snowflakes, resulting in the symmetric pattern shown in (B). (Wolfram Demonstrations: http://demonstrations.wolfram.com/3DSnowflakeFractals/)

a unidirectional manner. The universe expands and irrevocably changes the curvature of space and time, breaking the existing symmetry.

Symmetry and symmetry breaking in nature inspired the major foundation of 20th-century physics. It was Pierre Curie who first stated the importance of

Figure 1.3. Geometric symmetries are present in crystals (left). In quasicrystals the symmetric patterns are not repeated (right). (Wolfram Demonstrations: http://demonstrations.wolfram.com/Quasicrystals/)

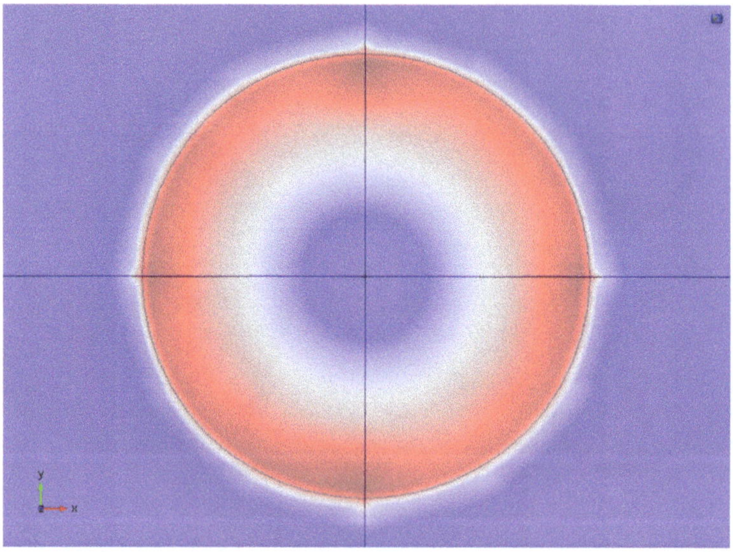

Figure 1.4. Symmetrical voltage contours around a test charge. (Comsol Multiphysics Simulation)

symmetry: 'When certain causes produce certain effects, the elements of the symmetry of the causes must be found in the effects produced' [1]. In other words, Curie proposed that in a physical phenomenon, the elements of the symmetry of its cause are also present in the resulting effect. In a further argument, he proposed that the asymmetry associated with an effect is also present in the cause. The generality of the statement made by Curie lost its value in science, where a formal mathematical definition or related models are considered important. However, it cannot be denied that Curie's ideas cast their reflections on future work in the field of symmetry and symmetry breaking.

In 1915, Emmy Noether proposed the relationship between symmetries in physical systems and the conservation of some associated physical quantities, which came to be widely regarded as the Noether theorem [2]. Put simply, the Noether theorem states that for every symmetry there is a conserved physical quantity. For example, in mechanics, the symmetry or invariance of energy with the passage of time is associated with conservation of energy. Similarly, the symmetry or invariance of the system's Lagrangian is associated with conservation of momentum. In electrodynamic systems, the symmetry of the electromagnetic field is expressed through gauge invariance which is associated with charge conservation [3]. These ideas have wide-ranging implications and deep ramifications for all the branches of physics. Today, it is widely believed that the quest for new laws and phenomena in physics is basically the quest for symmetries.

The study of symmetries in physical systems evolved over decades, leading to the discovery of the phenomenon of spontaneous symmetry breaking in the 1960s. The initial work was done by Yoichiro Nambu in the context of condensed matter physics, where he pointed out that the formation of Cooper pairs in superconductors is a case of spontaneous symmetry breaking—of the corresponding electromagnetic gauge symmetry [4]. The concept was put into a rigorous theoretical framework using scalar functions by Jeffrey Goldstone, a professor of physics at the University of Cambridge, who first proposed that for every spontaneous symmetry breaking, massless and energyless particles with even spin (called bosons) would be generated [5, 6]. These bosons are called Nambu–Goldstone bosons. Although these bosons have never been observed in accordance with the initial prediction by Goldstone in high energy physics, the theorem, with some modifications, has nonetheless been applied in various contexts. For example, Peter Higgs postulated that with spontaneous symmetry breaking of the local symmetry, the Goldstone bosons would acquire mass, resulting in massive particles; these came to be called the Higgs boson [7] or God particle, and were discovered in 2012 [8]. It is worth mentioning that Higgs had met Goldstone during a brief fellowship at CERN in 1965 and became fascinated by the concept of spontaneous symmetry breaking. Perhaps as a tribute to and an acknowledgement of the work of Jeffrey Goldsone, he named his Nobel prize lecture in 2012, 'Evading the Goldstone theorem'.

A related form of symmetry breaking is called explicit symmetry breaking, where the symmetry is broken in an explicit manner by external forces. Here, the ground state of the system in symmetry breaking is unique and the system's dynamic equations are not invariant [9]. Recently, a close relationship between electromagnetic radiation and the explicit symmetry breaking of electrodynamic systems was discovered [10]. In electrostatics, isolated charges have electric lines of field directed away from the centre of the charge in radial directions [11]. When the charge is briefly accelerated, the existing symmetry of the electric lines of force is explicitly broken.

The result of the acceleration is that the electric lines of field lose their rotational or spatial as well as temporal invariance. This symmetry breaking of the electric field under acceleration results in electromagnetic radiation.

Radiation from an electrodynamic system under explicit symmetry breaking, which is the central thesis of the book, offers a completely new perspective and a unified approach towards understanding the mechanism of radiation from diverse kinds of antennas—dielectric, piezoelectric, metallic wire or simple inductor–capacitor.

The concept of radiation under explicit symmetry breaking also helps in understanding the mechanism of electrodynamic systems, where, despite the radiation of electrons, there is no electromagnetic radiation, which is cancelled as a result of the symmetry of the system. A two-wire transmission line is one such case, where excitation fed by a time-varying voltage source does not result in electromagnetic wave emission despite electron acceleration, as the electric field lines have symmetry, i.e. invariance, along the spatial direction between the transmission lines. But when the wires are flared or opened up, there is a change in the shape of the electric field lines, i.e. symmetry is explicitly broken, resulting in radiation with an associated non-conserved current, which is emitted from the system.

The idea of explicit symmetry breaking found inspiration from some early work performed in the field of antennas. During early wireless technology experiments in 1895 at Salvan, Switzerland, Guglielmo Marconi found that by raising the transmitting wire vertically on the ground, the sensitivity of the system increases. In Italian, a tent pole is called *l'antenna centrale*. Marconi used the Italian word for pole to describe the radio transmitter and the word found wide use after the success of his transatlantic experiments on wireless communication. In Marconi's experiments, the wire or antenna was fed along only one of its terminals, and the other terminal was left floating in free space. The configuration allowed explicit symmetry breaking of the electric field, which resulted in radiation.

The concept can also be used to understand the phenomenon of radiation from dielectric resonator antennas, whose mechanism of operation was not clear until recently. The existing mathematical models talk about the operation of a dielectric resonator antenna from the perspective of a leaky waveguide. According to this model, a time-varying voltage source feeds the waveguide, which results in excitation of the electromagnetic modes in the waveguide, leading to leakage of radiation. The key problem with this model is that the fundamental issue of the transformation of time-varying current into electromagnetic modes is completely neglected. A leaky waveguide like a horn antenna does not radiate electromagnetic waves under time-varying excitation. It needs a feeding antenna, which transforms the input excitation into electromagnetic radiation, which leaks out of the system.

The operation of dielectric antennas can be understood from the perspective of explicit symmetry breaking of the electric field. A dielectric antenna mounted on a ground plane has a finite value of capacitance. The ground plane has some inductance. When electrical excitation is fed to the dielectric antenna on a ground plane, at a fundamental level, power is fed to a capacitor–inductor circuit in an asymmetric manner, which results in electromagnetic radiation. The concept can also be applied in the context of radiation from piezoelectric material-based thin film antennas.

Most electronic devices have a finite ground plane, which provides a common reference ground for voltage in them. No electronic device with a common ground plane is possible in practice. So, the concept of asymmetric excitation to capacitive antennas can be leveraged to design antennas which can be integrated with general electronic devices like television sets, mobile phones and LAN devices. As capacitive elements can be made using conventional semiconductor fabrication technologies, the idea can be used to develop antennas on a chip. It is worth mentioning that this would not result in miniaturisation of the entire antenna system. However, miniaturisation of the capacitive element of an antenna could certainly result in an overall size reduction.

Symmetry expresses itself along multiple dimensions of the convoluted and complex matrix of nature's expressions. It is inherently present in animate as well as inanimate objects. A more detailed investigation of the subtle aspects of symmetry and its transformations under different conditions will be pursued in the next couple of chapters.

References

[1] Curie P 1894 *J. de Phys.* **3** 401
[2] Noether E 1918 *Nachr. D. König. Gesellsch. D. Wiss. Zu Göttingen, Math-phys. Klasse* 235–57
[3] Quigg C 2013 *Gauge Theories of the Strong, Weak, and Electromagnetic Interactions* (Princeton, NJ: Princeton University Press)
[4] Nambu Y 1960 Quasi-particles and gauge invariance in the theory of superconductivity *Phys. Rev.* **117** 648
[5] Goldstone J 1961 *Nuovo Cimento* **19** 154
[6] Goldstone J and Salam A 1962 *Phys. Rev.* **127** 965
[7] Higgs P 1966 *Phys. Rev.* **145** 1156
[8] Chatrchyan S *et al* 2012 *Phys. Lett.* B **716** 30–61
[9] Brading K and Castellani E 2003 *Symmetries in Physics: Philosophical Reflections* (Cambridge: Cambridge University Press)
[10] Sinha D and Amaratunga G A J 2015 *Phys. Rev. Lett.* **114** 147701
[11] Jackson J D 1962 *Classical Electrodynamics* vol 3 (New York: Wiley)

IOP Publishing

Explicit Symmetry Breaking in Electrodynamic Systems and Electromagnetic Radiation (Second Edition)

Dhiraj Sinha and Gehan A J Amaratunga

Chapter 2

Symmetries and conservation theorems

2.1 Symmetry: a brief historical introduction

The word symmetry originates from the Greek word *symmetria*, which essentially means agreement in dimensions. *Symmetria* was first used in a treatise on sculpture, the *Canon* of Polyclitus, where it implied commensurability [1]. It was postulated that the key to achieving beauty and artistic perfection lies in the mutual commensurability of the constituent elements of a statue, as well as the complete structure. These ideas were further elaborated by Galen, a 2nd-century Greek physician and philosopher based in Rome, in support of Polyclitus:

> Beauty, Chrysippus [a Stoic philosopher] feels, resides not in the commensurability (*symmetria*) of the constituents (i.e. of the body), but in the commensurability of parts, such as the finger to the finger, and of all the fingers to the metacarpus and the wrist (carpus), and of these to the forearm, and of the forearm to the arm, in fact of everything to everything, as it is written in the *Canon* of Polyclitus. For having taught us in that treatise all the *symmetriae* of the body, Polyclitus supported his treatise with a work, having made a statue of a man according to the tenets of his treatise, and having called the statue itself, like the treatise, the *Canon* [2].

The meaning of the word symmetry went through a process of transformation from an aesthetic angle to an architectural one in the 18th century, when it came to be associated with images of structural balance in buildings, and then it further evolved into its modern association with the invariance of mathematical quantities [3]. A detailed discourse on symmetry was presented in a book by Vitruvius published in the 1st century BCE [4]. According to Vitruvius, 'symmetry is the appropriate agreement of the elements of the work itself, a correspondence [*responsus*], in any

doi:10.1088/978-0-7503-5132-4ch2

given part, of the separate parts to the entire figure as a whole. Just as in the human body, there is a symmetric quality of eurhythmies [*symmetros est eurythmiae qualitas*] expressed in terms of the cubit, foot, palm, digit, and other small units, so it is in perfect works [of architecture]' [4].

Claude Perraut, a French scholar, was commissioned by Jean Baptiste Colbert to translate *De Architectura* into French so that a new observatory could be constructed. In the process, Perraut gave a new meaning to symmetry. Perraut wrote:

> For our symmetry is properly the equality and the parity which is found between opposite parts, that is, if one makes, for example, an eye higher or larger than the other, or columns closer together to the right than to the left, or [their] number or size is not the same [to the right and to the left], one says that it is a defect of symmetry according to us [à nostre mode]: rather than, if a capital [of a column] is larger or a cornice protrudes more than the rules of order for a column allow, this is a defect of symmetry according to the ancients [5].

He further wrote [5]:

> Symmetry, although a word commonly used, does not signifie in French what Vitruvius understands by Proportion; for he understands by Proportion, a Relation according to Reason; and Symmetry, in French, signifies only, a Relation of Parity and Equality. For the word *Simmetria* signifies in Latin and Greek Relation only. As for Example, as the Relation that Windows of Eight Foot high, have with other Windows of Six Foot, when the one are Four Foot broad, and the other Three: and Symmetry, in French, signifies the Relation, for Example, That Windows have one to another, when they are all of an equal height and equal breadth; and that their Number and Distances are equal to the Right and the Left; so that if the distances be unequal of one side, the like inequality is to be found in the other.

Thus, Perrault defined symmetry in terms of the correspondence between two entities with respect to some axis. It gave a new meaning to the usage and interpretation of symmetry, one that is close to the modern comprehension and interpretation.

The initial use of symmetries in architecture continued to be reflected in ancient, medieval and modern structures. Figures 2.1–2.3 illustrate structures inspired by symmetries from ancient, medieval and modern times. In addition to enhancing the aesthetic value of structures, symmetries also increase the dynamic stability of the system and increase its longevity. Figure 2.4 is one such instance of symmetry in the design of modern bridges which are arch shaped. Many of the large symmetric structures have survived the ravages of time and events like earthquakes, floods and typhoons because of the ability of a symmetric structure to distribute stress. The concept is also used in simple designs, for example for manholes, which are round in shape. It is also reflected in the window design of ships (figure 2.5) and elevators (figure 2.6).

Figure 2.1. Roman amphitheatre. (Photo by DAVID ILIFF. License: CC-BY-SA 3.0)

Figure 2.2. Taj Mahal, built in medieval times represents an excellent masterpiece of architectural symmetry. (Photo by D Sinha)

The concept of symmetry in the architecture of Vitruvius, also had an impact in art and the most notable instance is L'Uomo Vitruvian or the Vitruvian Man, a drawing in pen and ink by Leonardo da Vinci which dates to 1490 and is kept in the Gallerie dell'Accademia of Venice in Italy [6]. It portrays a superposition of a man's figure in two different positions with arms and legs stretched out (figure 2.7). The whole image is inscribed in a circle and square. The drawing and accompanying text from the work by Vitruvius are referred to as the Canon of proportions.

Figure 2.3. Structures like the Sydney Opera House incorporate symmetry to illustrate art. (Photo by D Sinha).

Figure 2.4. Symmetries have engineering significance as they can provide stability and durability. (Photo by D Sinha).

As the drawing is inspired by *De Achitectura* of Vitruvius where the correlations between geometric proportions in human figure are described, it is called L'Uomo Vitruvian in his honour. Leonardo's drawing blends threads of knowledge from art and science at a critical time of the Renaissance and reflects his attempt to relate man to nature. The subsequent work of the artist closely reflected this philosophy

Figure 2.5. The windows of ships have a circular shape in order to provide durability. (Photo by D Sinha).

Figure 2.6. Symmetries are also incorporated in the design of multistoried structures like elevators, where stability is of special importance. (Photo by D Sinha).

which was adopted by other artists the world over, most notably Cesare Cesariano, Albrecht Dürer, Heinrich Lautensach and William Blake among others.

2.2 Symmetry in science

The projection of the concept of symmetry into the field of mathematics, along the dimensions of its current narrative, was done by the French mathematician, Adrien-Marie Legendre. Legendre defined symmetry angles in his book, *Elements of Geometry*: 'two equal solid angles which are formed (by the same plane angles) but in the inverse order will be called angles equal by symmetry, or simply symmetric angles' [7].

The modern conception of symmetry is associated with a property that leads to its invariance under a set of physical transformations, such as translation, rotation,

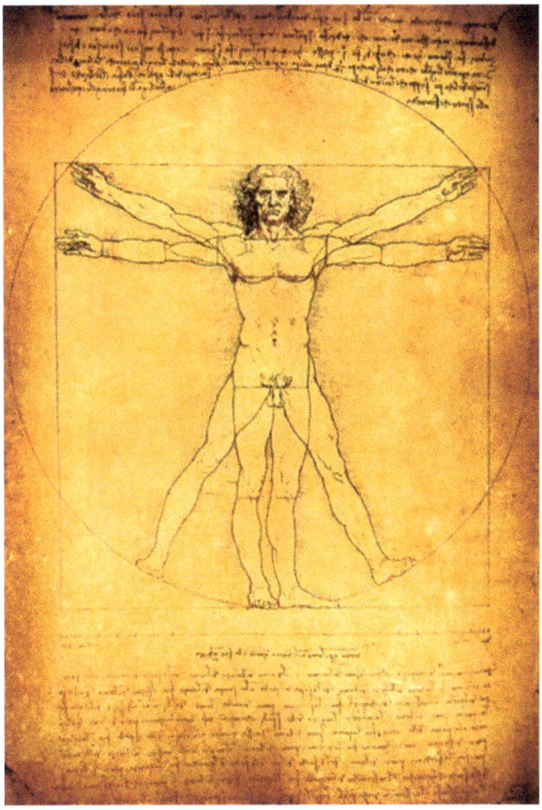

Figure 2.7. Vitruvian Man by da Vinci. Reproduced from https://commons.wikimedia.org/wiki/File%3ADa_Vinci_Vitruve_Luc_Viatour.jpg, Leonardo da Vinci [Public domain], via Wikimedia Commons.

reflection or inversion. This modern narrative of symmetry is described by Hermann Weyl in his book, *Symmetry*, where he associates symmetry with balance and proportion among a set of entities which are integrated into a large whole, while correlating it with beauty, but he also emphasises the fact that symmetry is additionally associated with operations like reflection or rotation, which lead to spatial invariance [8].

The symmetry of a mathematical entity is expressed through its invariance under the transformation of a specific variable related to the given entity, which could be temporal, i.e. related to the passage of time, or spatial, i.e. related to a geometric transformation, such as rotation, translation, reflection, division or scaling [9]. An example of rotational symmetry is a perfect sphere whose rotation in three-dimensional space leads to invariance of its physical shape. Another example of such a symmetry is a circle whose physical shape remains invariant under rotation in two-dimensional space. Similarly, a line has a translational symmetry and as we move along it, the linear points remain invariant or indistinguishable from each other. These are also examples of reflectional symmetry, as the objects have a central axis along which reflection leads to invariance of shape.

Symmetries are not absolute but relative and the order of symmetries can be changed with transformation. For example, a solid sphere having rotational symmetry in three-dimensional space can be transformed into a disc that has two-dimensional rotational symmetry. Similarly, a disc can be transformed into a line that has one-dimensional symmetry. Such acts of transformation where the symmetry is lowered are called symmetry breaking. So, during each phase of transformation, symmetry is not lost completely but transformed into a different order. Symmetry is also used in logic, where a relation is said to be symmetric. For example, if a is equal to b, then b is also equal to a. It is also widely used in group theory, where the set of mathematical operations which result in the invariance of a given variable of a mathematical entity are said to form a group. For example, the ratio of the circumference and diameter of a circle remains invariant to changes in their values.

In one of the biggest discoveries of 20th century physics, Albert Einstein considered symmetry as the fundamental attribute of nature which defines the dynamics of physical laws. He argued that electromagnetic fields were transformed as a consequence of relativistic invariance [10]. He identified the symmetry present in Maxwell's equations and applied it in the context of space–time fabric of the universe. So, in a way, he created geometric framework around nature's symmetry. This was particularly evident in his general theory of relativity where the invariance of the laws of nature with respect to changes in localized coordinate systems defined the nature of gravity.

In physics, symmetry specifically means the invariance of a physical quantity like energy or momentum with transformation of one of the temporal or spatial dimensions. In a landmark work, the relationship between symmetries and conservation was first observed by Emmy Noether, perhaps the greatest female mathematician of all time [11]. Her initial work was in the field of invariants in algebra and dealt with the features of mathematical functions which remain unchanged when they are transformed. Noether was invited by David Hilbert, one of the most notable mathematicians of all time, to take lectures at the University of Göttingen. In 1915, shortly after the publication of Einstein's paper on the general theory of relativity, Hilbert asked Noether to apply her work on algebraic invariants to the equations of general relativity. Noether made the surprising discovery that for every symmetry in a physical system, there is a conserved quantity that is also called the conserved current, or Noether current [12]. This is reflected in the time-based or temporal symmetries of physical systems. For example, the total energy of an isolated system remains the same with time or shows temporal invariance or symmetry. This is associated with the conservation of energy. In this context, the conserved current is the energy of the system. Another instance of symmetry is reflected in spatial invariance. When two bodies collide, the force exerted by the first body on the other is equal and opposite to the force exerted by the second body on the first. The net change in force on the system comprising the two bodies remains unchanged during the collision. In other words, the force remains invariant, or there is symmetry. This symmetry is associated with the conservation of momentum.

Noether's work brought symmetries to the heart of scientific endeavour and made an impact in all the domains of physics. In electromagnetism, the symmetry of the electromagnetic field is expressed using gauge invariance, which is associated with the global conservation of charges and currents [13].

However, it must be stated that not all symmetric systems have associated conserved quantities. The relationship between symmetries and conservation applies mainly in the context of systems with conservative force fields. In dissipative systems with frictional or drag forces, despite the inherent symmetries, conservation laws may not necessarily hold.

2.3 Symmetries in dynamic systems

Symmetries in art and geometry are evident through structural forms. In dynamic systems, the interactions are represented through a set of mathematical formalism widely referred to as the Lagrangian formulation which offers the freedom to define the dynamics while expressing the interactions using new coordinate systems like position and momentum. It is worth mentioning that in Newtonian formulation of dynamical systems, the position coordinate is an essential component which can become cumbersome in terms of expressing the interactions of complex systems.

2.3.1 Lagrangian formulation

The central idea of the Noether theorem is built around the invariance of the Lagrangian of a dynamic system with a change to one of its variables and the conservation of some corresponding physical quantity. The Lagrangian is a mathematical function that represents the dynamics of a system. In mathematical terms, it is the difference between the kinetic and potential energy of a system [14]. The values of the Lagrangian of a system at each instant in time can be used to calculate the trajectory of a particle. The area under the curve between the function representing the Lagrangian and time gives the value of a physical quantity called action which holds the least value for a given dynamic system, which is also called the stationary value of the given integral. Any other path between the initial and final values of the Lagrangian leads to a larger action. It is a mathematical portrayal of the principle of least action, or the fact that events happen in order to minimize the action.

The Lagrangian formulation allows the use of variables other than the conventional coordinate system to represent the dynamics of a system. These variables are called generalised coordinates and they enable physical constraints to be incorporated into the description of the physical state. The importance of the Lagrangian in physics is expressed by Robert Adair in *The Great Design: Particles, Fields and Creation*: 'Indeed, at the present time it appears that we can describe all the fundamental forces in terms of a Lagrangian. The search for Nature's One Equation, which rules all of the Universe, has been largely a search for an adequate Lagrangian' [15].

If the kinetic energy of a system is T and the potential energy is U, the Lagrangian is given by $L = T - U$ [14]. The generalised coordinates of a physical quantity are given by $\{q_1, q_2, q_3 \ldots\ldots q_n\}$ where n is an integer and their corresponding derivatives with respect to time are expressed as $\{\dot{q}_1, \dot{q}_2, \dot{q}_3 \ldots\ldots \ldots \dot{q}_n\}$. To cite a simple example, the Lagrangian of a body of mass m with velocity $\dot{q}$ thrown at a height h is given by $1/2m\dot{q}^2 - mgh$. A dynamic system's evolution is described by an action integral, which is given by [14]

$$\mathcal{L} = \int_0^T L(q_1, q_2 \ldots\ldots \dot{q}_n, \dot{q}_1, \dot{q}_2 \ldots\ldots \dot{q}_n)\,\mathrm{d}t. \tag{2.1}$$

Here, the Lagrangian is integrated from time $t = 0$ to $t = T$. Equation (2.1) has stationary values, which implies that $\mathrm{d}\mathcal{L} = 0$. By making related substitutions and using the principles of calculus of variation, we obtain the Euler–Lagrange equation given by,

$$\frac{\mathrm{d}L}{\mathrm{d}q_i} = \frac{\mathrm{d}}{\mathrm{d}t}\left(\frac{\partial L}{\partial \dot{q}_i}\right). \tag{2.2}$$

Here, the generalized coordinates vary over a set of integers $i = 1, 2 \ldots n$. The solution of equation (2.2) is equivalent to finding a trajectory of the particle for which the action assumes stationary values. The left-hand term of equation (2.2) is the derivative of the Lagrangian with spatial coordinates and represents the force exerted on the system. The right-hand side of equation (2.2) represents the rate of change of the momentum or force. Thus, equation (2.2) is an alternative expression of Newton's second law of motion, which relates force to the rate of change of the momentum. It can be applied to calculate force in simple dynamic systems. For example, the force on a body of mass m, velocity $\dot{q}$ at a height Z in a gravitational force field is mg and it is equal to the derivative of the Lagrangian with spatial coordinate $mg = \mathrm{d}L/\mathrm{d}q$. The momentum is given by mv and is equal to the $\mathrm{d}L/\mathrm{d}\dot{q}$. Thus, the Euler–Lagrange equation basically equates force to the time rate of the change of momentum. Its further simplification can lead to the development of the trajectory of a particle in a given force field. In the context of a spring mass system where the spring is compressed by x and the spring constant is k, the Lagrangian's value is $1/2m\dot{q}^2 - 1/2kq^2$. The force is given by $\mathrm{d}L/\mathrm{d}q = kq$ and the momentum is given by $\mathrm{d}L/\mathrm{d}\dot{q}$, whose rate of change equals the gradient of the potential energy ΔU.

The Lagrangian for a body of mass m and charge e in an electromagnetic field is given by

$$L = \frac{1}{2}m\dot{q}^2 + \dot{q}eA - eV, \tag{2.3}$$

where A is the magnetic vector potential and V is the electrical potential. The momentum is expressed by $\mathrm{d}L/\mathrm{d}\dot{q} = m\dot{q} + \dot{q}eA$.

By replacing the Lagrangian in equation (2.1) with a general function and integrating it with one of the corresponding variables, we can obtain the foundational principle of the calculus of variation, which is used to calculate the shortest distance between two points on a surface, the path of least descent of a particle, the area generated by the rotation of a curve in free space, and related problems.

The main advantage of the Euler–Lagrange equation is that the constraint forces can be avoided by considering generalised coordinates such as angle, momentum or distance, which define the system's number of degrees of freedom. The generalised coordinates are generated from the constraints of the system, which are smaller in number than the total number of coordinates, hence, as the number of constraints increases, the Newtonian approach becomes complex but the Lagrangian approach becomes simpler. Another related advantage is that the Lagrangian formulation has the same form in mechanics, hydrodynamics and electrodynamics, as well as quantum field theory and general relativity, and the problem is expressed in terms of potential and kinetic energy, which is simpler in comparison to forces, which tend to have different geometric orientations.

Another key advantage of the Lagrangian approach is that it opens up new perspectives regarding analysis of the system in terms of its symmetries, which are discussed in the following section.

2.3.2 Noether theorem

The fundamental idea of Noether's theorem is that symmetries in physical systems are associated with conservation of some associated physical quantity [11, 12]. As discussed earlier, symmetry implies the invariance of the state of a physical system, geometric entity or mathematical function with some corresponding transformation. In more specific terms, Noether's theorem postulates that the symmetry or invariance of the action of a physical system which is the integral of Lagrangian with time is associated with a conserved quantity. The conservation laws are always applied in the context of closed systems. In simple terms, a closed system is a collection of entities on which outside forces are zero. Systems can never be closed in the perfect sense, however, if the impact of external forces is negligible, a system can be considered to be closed from a practical perspective.

A very simple example of Noether's theorem is the invariance of force in a dynamic system which is associated with conservation of momentum. When a bat strikes a ball with a force F, the ball exerts an equal and opposite amount of force on the bat. The sum of the two forces is equal to zero. This implies that the net force on the system before and after the collision remain the same. Force is the differential of the Lagrangian of the dynamic system with the position coordinate, which indicates the invariance of the Lagrangian in the spatial dimension or its symmetry. This also means that the net change in momentum remains zero, i.e. the initial momentum is equal to the final momentum. This is essentially the law of conservation of momentum. Thus, the invariance of the Lagrangian with position, or the translational symmetry of the Lagrangian, is associated with momentum conservation.

The relationship between the translational symmetry of the Lagrangian and momentum conservation is universal and applies to all dynamic systems. In certain cases, the momentum conservation may not be that apparent. For example, a body of mass m is dropped from a certain height. We assume that the dissipative frictional forces are negligible. As it travels towards the surface of the Earth, its potential energy is transformed into kinetic energy and it gains a net momentum. The positive value of the momentum might lead one to conclude that momentum is not conserved. The fact is that the ball is being pulled by the Earth with a force F, but it also pulls the Earth towards itself with the same force in the opposite direction. So, the net force on the ball and the Earth comes out to be zero, which essentially means that the value of the change in the Lagrangian with position coordinate $\mathrm{d}L/\mathrm{d}q$ is zero, or the Lagrangian has translational symmetry. The initial momentum of the ball is zero and as it moves towards the surface of the Earth, it gains a finite momentum mv. In the process, the Earth with mass M is also pulled by the ball and moves towards it with a velocity V and in the process, $mv+MV=0$, which is the law of conservation of momentum. Thus, the invariance of Lagrangian along the spatial dimension is associated with conservation of momentum.

There are other instances of symmetries and conservation. For example, in rotational systems, the Lagrangian shows rotational invariance, as it is not dependent on the angular coordinate. This corresponds to conservation of angular momentum. A very simple illustration of this phenomenon is the rotational motion of an ice skater who is able to increase the rotational speed by folding her arms. An interesting consequence of conservation of angular momentum is observed in the Earth–Moon system, where the angular momentum of the Earth on the Moon results in the reduction of its rotational speed at the rate of 65.7 nanoseconds per day, which causes an increase in the radius of the Moon's orbit of 3.82 cm yr^{-1}.

Another interesting aspect of the Noether theorem is observed in the temporal invariance of the Lagrangian. This means that the derivative of the Lagrangian with time is equal to zero, i.e. $\mathrm{d}L/\mathrm{d}t=0$. This invariance is associated with conservation of energy. In dynamic systems, energy is neither lost nor created, which is reflected in the constant value of the total Hamiltonian, which is also related to the Lagrangian. This invariance is associated with conservation of energy.

An important consequence of Noether's Theorem is that if a physical system or a process shows invariance of its Lagrangian with time or space, the inherent symmetry implies conservation of momentum and energy. Thus, the ability to identify some conserved quantities from the symmetries of a system is one of the greatest achievements of Noether's theorem, which has radically influenced scientific advancements.

The Noether theorem can be expressed in simple terms by considering the invariance relationship between two different physical quantities. For example a quantity Y is said to be invariant with respect to X, if variation in X does not result in any variation in Y. Mathematically, it can be expressed as,

$$\frac{\mathrm{d}Y}{\mathrm{d}X} = 0. \tag{2.4}$$

Here, Y is also said to be a conserved quantity.

In a dynamic system, such quantities are called constants of motion. Historically, momentum and energy were the initial physical quantities that were regarded as constants of motion. Interest in the field of invariants increased with the development of Lagrangian mechanics based on the principle of least action, which incorporated the concept of generalised coordinates. It states that the physical trajectory adopted by a system during motion is such that the value of action is constant, which leads to the Euler–Lagrange equation. In the case of closed systems, where external forces are zero,

$$\frac{\mathrm{d}L}{\mathrm{d}q_i} = \frac{\mathrm{d}}{\mathrm{d}t}\left(\frac{\partial L}{\partial \dot{q}_i}\right) = 0 \tag{2.5}$$

and

$$\frac{\mathrm{d}}{\mathrm{d}t}\left(\frac{\partial L}{\partial \dot{q}_i}\right) = \frac{\mathrm{d}p_k}{\mathrm{d}t}, \tag{2.6}$$

where $\partial L/\partial \dot{q}_i = p_k$ is conserved during the motion over a set of values of $k = 1, 2 \ldots n$. Here, the Lagrangian remains invariant to changes in the coordinate system, which means it has translational symmetry and is associated with momentum conservation. To be more specific, we consider the invariance of action under perturbation of time and generalized coordinates which can be expressed as, $t \to t + \delta t$ and $q \to t + \delta q$ where δt and δq are small variations in time and generalized coordinates, respectively. If there are N symmetric transformations of the action integral expressed by indices, $j = 1, 2, 3 .., N$, then the resulting perturbation can be expressed as a linear combination of the individual perturbations, $\delta t = \sum_j \varepsilon_j T_j$ and $\delta q = \sum_j \varepsilon_j Q_j$. Here ε_j indicates parametric coefficients, T_j is the generator of time evolution and Q_j is generator of the generalized coordinates. These expressions imply that there are N conserved quantities expressed by [12],

$$\left(\frac{\partial L}{\partial \dot{q}}\dot{q} - L\right) \cdot T_j - \frac{\partial L}{\partial \dot{q}} Q_j = 0. \tag{2.7}$$

The Lagrangian has three key symmetries: translational, rotational and temporal. If there is a Lagrangian that is independent of time or, in other words, it has temporal symmetry, it retains its initial value with a change in time from t to $t + \delta t$. Under such a condition, $N = 1$, $T = 1$ and $Q = 0$, so, the conserved quantity is the total energy of the Hamiltonian of the system given by,

$$H = \frac{\partial L}{\partial \dot{q}}\dot{q} - L. \tag{2.8}$$

In a closed system, the external force is zero and the Lagrangian does not depend on coordinate q_k. The Lagrangian has translational symmetry and it retains its value under a change q_k to $q_k + \delta q_k$. Here, the values of the related variables are $N = 1$, $T = 0$ and $Q_k = 1$. The conserved quantity is momentum, $p_k = \partial L/\partial \dot{q}_i$.

In rotational dynamics, the Lagrangian has rotational symmetry, as it does not depend on the angular coordinate. With a small change in angle, the corresponding Cartesian coordinate r changes by $r + \delta\theta n \times r$ where $\delta\theta$ is an infinitesimally small change in angle θ and n is a coordinate axis. As time remains constant, we can consider $T = 0$, and the generalised coordinates q are comprised of the Cartesian coordinates r. Q is given by the following expression, $Q = n \times r$. The Noether theorem leads to the following conserved quantity,

$$\frac{\partial L}{\partial \dot{q}} \cdot Q_r = p \cdot (n \times r) = n \cdot (r \times p) = n \cdot L. \tag{2.9}$$

It implies that the angular momentum along the direction of the n axis remains the same or is conserved. For an arbitrary value of n, the system can be rotated along any direction, leading to the conservation of angular momentum.

As Noether theorem expresses symmetries in nature, the theorem must hold some kind of inherent symmetry. For example, its converse must also hold true. If a particular system undergoes some kind of change from an initial stage to a final stage such that a particular quantity loses its conserved value, it could be argued that the system's initial symmetry has undergone some kind of change. In other words, non-conserved physical quantities correspond to symmetry breaking. This aspect will be dealt with in some detail at a later stage in the book.

2.4 From symmetry to gauge theory

Study of physics is a theoretical and empirical examination of the relationship between physical quantities which vary. Early work in the field of calculus by Newton and Leibniz was a turning point of physics as the study of these variations was put in a formal manner. However, the work by Lagrange highlighted the role of quantities which do not vary under a certain set of mathematical operations. The action integral is one such physical quantity which has a stationary value for a given situation. The work by Noether further highlighted the role of invariants in physics. As invariants are related to symmetries, physicists like to formalise the physical phenomenon from the perspective of local and global symmetries.

A given physical situation can be expressed using different but equivalent mathematical formulations. For example, the motion of a body in space can be represented by different coordinate systems which are related to each other through Galilean transform. These different representations are said to form a group of symmetries which define the physics of a dynamic system [13]. The ideas of local and global symmetries are built around these different kinds of representations. A gauge theory is a simple example of such a mathematical model built around local and global symmetries through a related set of transformations.

As physical laws are defined in terms of coordinate systems, it is important to see the variation of these quantities with a transformation of the associated coordinate system. It turns out that physical quantities like position, electric charge, magnetic field etc retain their values with a transformation in the coordinate systems. These quantities are said to have global symmetries. Fields play a central role in understanding nature and natural phenomena. For example, the electric and magnetic fields are the central aspects of electromagnetism while the gravitational field determines the dynamics of celestial bodies. A common aspect of all fields is that certain quantities cannot be measured in absolute terms, for example, as mentioned earlier, the electric potential in electromagnetism cannot be measured in absolute terms. *However, there are certain transformations among the configuration of these unmeasurable quantities which can be measured and these transformations are called gauge transformations.* During the transformation of these quantities from one configuration to the other, there are certain variables which do not change and this property is called gauge invariance. This invariance is also referred to as gauge symmetry and invariance is normally associated with some kind of symmetry. A theory of physics which possesses gauge invariance of some kind is called gauge theory.

The importance of gauge symmetry lies in the fact that it sets some kinds of limits on observables. As quantum mechanics is fully based on observables, the concept acquires some vital importance in this field. One of the most landmark aspects of 20th century physics is that forces emerge due to constraints associated with localised gauge symmetries. In quantum field theory, it is assumed that the constraints of gauge symmetry result in forces which are described using particles called gauge bosons. This is the central aspect of the Standard Model which explains interaction between subatomic particles.

The basic ideas of gauge theory have emerged from the concept of gauge symmetry. At a fundamental level, gauge theory is built around the concept of invariance of the Lagrangian under a set of transformations. The word gauge mainly refers to the variables on which the Lagrangian is dependent. These are also called the number of degrees of freedom of the Lagrangian. The set of functions which is obtained through transformation between these variables or gauges is called the symmetry group. They are also called the Lie group or the gauge group. Any transformation of the Lagrangian which leads to one of the given Lie groups is called gauge field. As mentioned earlier, the Lagrangian remains invariant to the local transformations, which is called gauge invariance. The gauge fields ensure invariance of the Lagrangian. The quantised forms of gauge fields are called gauge bosons. In most cases, the symmetry group is commutative, however, in certain cases, it is non-commutative and then the gauge theory is called non-Abelian. The Yang–Mills theory is one such example of a non Abelian gauge theory.

If the Lagrangians are invariant under a transformation which has been done continuously at every point of space in the domain of the physical process, the Lagrangian is said to have a global symmetry. However, if the transformations are carried out at some specialised points in the limits of space of the physical process resulting in invariance, the Lagrangian is said to have local symmetry.

2.4.1 Gauge symmetry

Symmetries in electromagnetism are expressed through gauge symmetry. The Lagrangian of an electromagnetic field is dependent on electric and magnetic field intensities. Invariance of the Lagrangian implies invariance with respect to change of one of these quantities. It turns out that the electric and magnetic field intensities are related to magnetic vector potential and the electromagnetic Lagrangian satisfies the Noether theorem with a set of conserved quantities.

In gauge symmetry, a modified form of Noether theorem called Noether's second theorem is used which generalises the relationship between symmetries and conservation laws. According to it, if the action, i.e. the integral of the Lagrangian with time, has infinitesimal symmetries defined by a number of functions N along with their derivatives up to a certain order, then the derivatives of the Lagrangian satisfy N number of differential equations. Thus, there is a one-to-one correspondence between the conserved currents and the associated symmetries of the Lagrangian which are called gauge symmetries. It could be easily seen that it points towards the degeneracy of the Lagrangian, i.e., it indicates that a number of Lagrangians having the same numerical values are possible. In gauge theory, the conserved quantities are called Noether identities and they are associated with the degenerate states of a Lagrangian. In mathematical terms, the Noether identities can be defined in terms of a differential operator. The kernel of the given differential operator has a range of Euler–Lagrange operators related to the given Lagrangian.

The concept of gauge symmetry or gauge invariance evolved from classical electromagnetism where the electric and magnetic potentials cannot be measured in absolute terms. For example, the electric potential, is always measured with respect to a ground potential which is assumed to be present at a point infinitely away from a test charge. This is considered to be an unobservable quantity, however, the potential difference between two points can always be measured, which is some kind of transformation between two configurations of the potentials. The electric field is the gradient of the electric potential which means that the electric potential can be found by integrating the electric field and adding a constant. The electric potential V differs by a constant C, i.e., $V \to V+C$ which means there can be more than one set of potentials that can give the same value of electric field. That is, the electric field is not a uniquely defined quantity and the same electric field can be associated with more than one potential function. *This means that the electric field can remain invariant if some explicit change is introduced in the electric potential.* A simple way to state this is that an electric field is invariant to changes in the corresponding potentials by a finite constant. The electric field is the gradient of the electric potential defined by $E = -\nabla V$ and while calculating electric field at any point, any constant term in the definition of electric potential can get cancelled out. This aspect of invariance is further broadened to incorporate invariance of the magnetic field. The magnetic field is related to a magnetic vector potential A, by the following relation, $B = \nabla \times A$. The general expression for electric field is given by, $E = -\nabla V - \partial A/\partial t$. The gauge transformation takes the form, $A \to A + \nabla f$, where f is a function of position and time and ∇f is its gradient, and $V \to V - \partial f/\partial t$. These equations indicate that the electromagnetic field vector remains the same, i.e., maintains an invariance with a change in transformation of the values of electric and magnetic potentials. These transformations are called gauge transformation and the corresponding invariance of the electromagnetic field is called gauge invariance. As the Lagrangian of an electromagnetic field is defined by the electric and magnetic fields, the Lagrangian also shows an invariance to these changes and the corresponding symmetry is reflected. A related statement which summarises these observations is the following—'Maxwell's equations follow gauge symmetry'.

2.4.2 Gauge theory

Although, the ideas related to gauge theory emerged in the context of electromagnetism and in the general theory of relativity, they have entered the field of quantum electrodynamics which is quantum mechanics merged with special theory of relativity. They are integral parts of the Standard Model, condensed matter physics and in recent years of string theory. An interesting aspect of gauge symmetry is that, although, the ideas were a part of Maxwell's equations published in 1864, physicists were a bit late to identify these until the first quarter of the 20th century. In the early years of General theory of relativity, David Hilbert provided an alternative derivation of the field equations by considering the invariance of action integral with coordinate transformation by using generalised variables. Hermann Weyl made an unsuccessful attempt towards unifying the general theory of relativity and electrodynamics while postulating the invariance of a special gauge. Later, he established the connection between gauge symmetry and global conservation of charges in electromagnetism [16]. The idea had profound impact in quantum field theory. Weyl is credited with the discovery of the concept of gauge symmetry.

Symmetry in quantum mechanics

A field theoretic version of Noether's theorem discusses the symmetries and conservation in a 4-dimensional space-time manifold expressed by x^μ where μ is an index which varies over the temporal and spatial coordinate variables. Such a formulation is widely used in modern physics. We assume a differentiable function ϕ representing a given field embedded within space–time manifold. It could be electric field intensity, magnetic field or a temperature distribution. Applying the principle of least action to it, we can write [13],

$$\Lambda = \int L\left(\varphi, \partial_\mu \varphi, x^\mu\right) \mathrm{d}^4 x \tag{2.10}$$

If the action is invariant under change of coordinates, we can make some substitutions, $x^\mu \to x^\mu + \delta x^\mu$ and $\varphi \to \varphi + \delta\varphi$. We can expand the tensors, $\delta x^\mu = \varepsilon_j X_j^\mu$ and $\delta\varphi = \varepsilon_j \psi_j$, where $j = 1,2,3.....,N$. Under such a condition, there are N conserved quantities according to Noether's theorem expressed by,

$$J_r^\nu = -\left(\frac{\partial L}{\partial \varphi_{,\nu}}\right) \cdot \Psi_r + \left[\left(\frac{\partial L}{\partial \varphi_{,\nu}}\right) \cdot \varphi_{,\sigma} - L\delta_\sigma^\nu\right] X_r^\sigma \tag{2.11}$$

Here, ν, σ and r, are indices which take integer values 0, 1, 2, 3.

The Noether current is expressed as $\partial_\mu J^\mu = 0$. The equation basically expresses the fact that a quantity is conserved within a given region of space and time unless it specifically leaves the system.

If a given system has symmetry, i.e. its Lagrangian $L\left(\varphi, \partial_\mu\varphi, x^\mu\right)$ does not change with a change in its variables. Here, the value of $N = 4$ as there are 4 variables on which the Lagrangian is dependent, out of which three are space variables and one is time. Here, the space-time variables are being changed and the field remains constant, hence, $\psi_r = 0$ which means $\delta\varphi$ is also zero. The value of the tensor variable representing the space-time coordinates X_j^ν equals δ_j^ν. By using these variables, the conserved current is given by

$$T_j^\nu = \left[\left(\frac{\partial L}{\partial \varphi_{,\nu}}\right) \cdot \varphi_{,\sigma} - L\delta_\sigma^\nu\right]\delta_j^\sigma = \left(\frac{\partial L}{\partial \varphi_{,\nu}}\right) \cdot \varphi_{,j} - L\delta_\mu^\nu \tag{2.12}$$

This is the stress energy tensor.

When we assume $X_j^\nu = 0$ while considering the variation of the field φ, the conserved quantity which we get is the charge. In quantum mechanics, the particle is described by a complex wavefunction, ψ, whose square gives the probability of finding a particle at a certain point in space. Thus, the field remains invariant to transformation the complex wavefunction to its complex form ψ^*. For example when ψ is replaced by $e^{i\theta}\psi$ or ψ^* is replaced by $e^{-i\theta}\psi^*$, the probability amplitude of the particle remains constant. The invariance under phase changes leads to the condition of charge conservation i.e. the total charge associated with the given quantum system remains constant.

For very small values of the phase, we get the following Lagrangian,

$$L = \psi_{,\nu}\psi^*_{,\mu}\eta^{\nu\mu} + m^2\psi\psi^* \tag{2.13}$$

The conserved current equals,

$$J^\nu = i\left(\frac{\partial\psi}{\partial x^\mu}\psi^* - \frac{\partial\psi^*}{\partial x^\mu}\psi\right)\eta^{\nu\mu} \tag{2.14}$$

The electric current density associated with the particle can be found by multiplying the current with that particular charge on the particle.

Group theory and symmetry

The fields and related symmetries in physics are described using a specific coordinate system whose dimensions have a specific relationship through the related gauge transformations. These coordinate variables are called fibre bundles. A group consists of a number of elements each of which have an inverse within the group and are associated to each other through the operation of multiplication [13]. For example, if a is an element of a group G, a^{-1} is also an element and there are elements b and c such that $a = b\,.\,c$. The group is called Abelian if it follows the law of commutation. The product of two groups is defined by pairs of elements defined by a multiplication rule. For example, the product of elements of two groups G with elements g_1 and g_2 and H with elements h_1 and h_2 can be represented by $(g_1, g_2).\,(h_1, h_2) = (g_1.g_2, h_1.h_2)$.

The set comprising all the possible finite transformations at some space-time point is called finite dimensional Lie group. In Gauge theory, a set of $n \times n$ unitary matrices, U(n) represent the symmetries of fields. Here, n represents the size of dimensions of the given matrix. The set of unitary matrices form a group and are referred to as SU(n) group. When $n = 1$, the group represents circles comprising complex numbers of unity amplitude. At a geometric level, U(1) represents the rotational symmetry of a circle. The fact is that, a circle of unit radius can be a simple representation of a complex wavefunction of an electron and any phase transformation would result in a new position along the unity circle. Thus, all the points along the circle represent a set of related numbers which can be described by the group. Because of this, symmetry of the electron wavefunction with phase changes is widely referred to as U(1) symmetry where U(1) represents the unity matrix which has a set of complex numbers as its elements. When quantum mechanics developed, the concept of U(1) gauge symmetry was developed by several scientists out of which Fock was a leading figure. U(1) gauge symmetry describes the symmetry of the wavefunction with a change in phase and it is able to describe the impact of an electromagnetic wave on an electron or a related quantum mechanical entity.

2.4.3 Symmetries in particle physics

One of the most fundamental building blocks of matter are leptons which are characterised by finite rest mass, charge and spin. The leptons are comprised of electrons, muons, tau and associated neutrinos [16]. The electron is the most stable among all these and the lightest. Muons have charge and spin equal to that of electrons but are heavier in comparison to electrons by a factor of 205.7 and have a mean lifetime of 2.2 μs. Tau particles are more massive than muons by a factor of 16.8 and have a mean lifetime of 0.29 picoseconds. The leptons interact through electromagnetic and gravitational forces and tend to have a corresponding antiparticle called antilepton which differs from the lepton in terms of some physical quantities having an opposite value. However, it is not yet established that neutrinos have their own antiparticle.

Other fundamental building blocks of matter are called hadrons, which are made of quarks which interact with each other through strong force. Hadrons belong to two families—baryons which are comprised of three quarks and mesons which have one quark and one antiquark. Protons and neutrons are stable forms of baryons and pion is an example of a meson. An important difference between baryons and mesons is that all baryons are fermions (they have fractional spin) while all mesons are bosons (they have integral spin).

Quarks, the fundamental building blocks of hadrons which were independently proposed by Murray Gell Mann and George Zweig in 1964, are never found in isolation and are either found in baryons or in mesons [17]. They have physical properties such as electric charge, mass and spin. Besides these, they also have additional properties called flavour and colour which have absolutely nothing to do with the linguistic meaning of 'flavour' and 'colour'. Quarks are said to possess six different flavours: up, down, strange, charm, top and bottom. They are classified into red, blue and green based on colour. All the quarks have an anti-particle called anti-quark. They experience electromagnetic, weak as well as strong forces and are the only elementary particles whose charge is not an integral multiple of the charge of an electron.

The heavier quarks decay into lighter quarks in high energy experiments. The lightest and the most stable of quarks are up and down quarks, each carrying a charge of 2/3e and −1/3e, respectively. They are not found in isolation. A neutron comprises of two down quarks and one up quark, similarly, a proton is comprised of two up quarks and a down quark.

The quarks and leptons belong to a broad class of particles called fermions which have half integral spins. A related class of particles called gauge bosons are the particles with spin 1 which mediate electromagnetic, strong and weak interactions. Examples of gauge bosons are photons, W bosons, Z bosons and gluons (figure 2.8). Photons are mediators of electromagnetic interactions and they interact with particles having nonzero electric charge. The electromagnetic force between photons is mediated through exchange of photons, as illustrated in figure 2.9. Gluons are a type of gauge bosons which mediate the strong interaction and are massless.

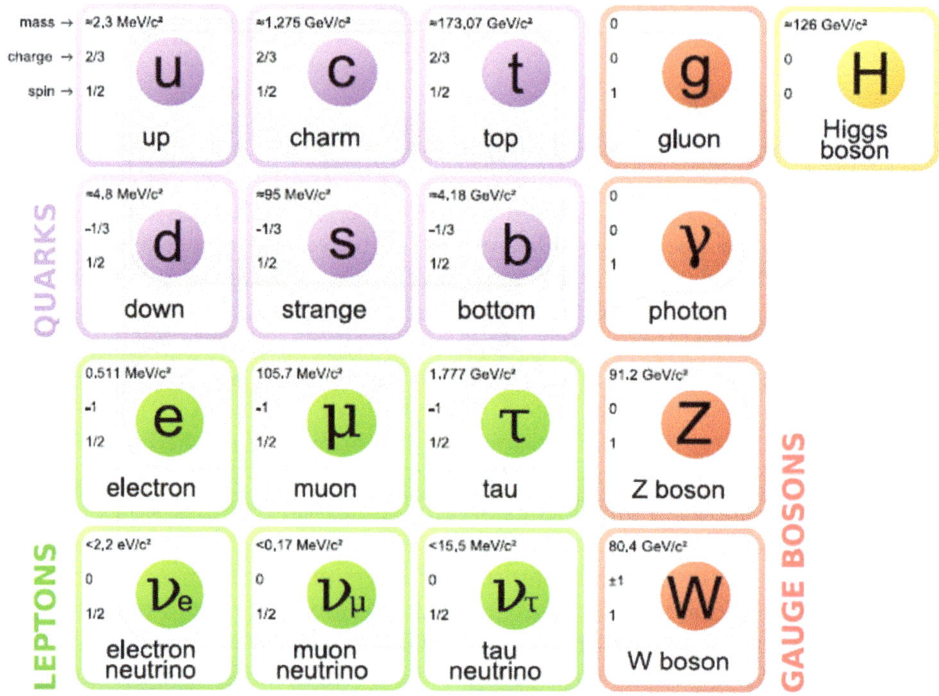

Figure 2.8. Pictorial representation of elementary particles. Reproduced from Wikipedia https://en.wikipedia.org/wiki/Standard_Model under the creative commons licence 3.0.

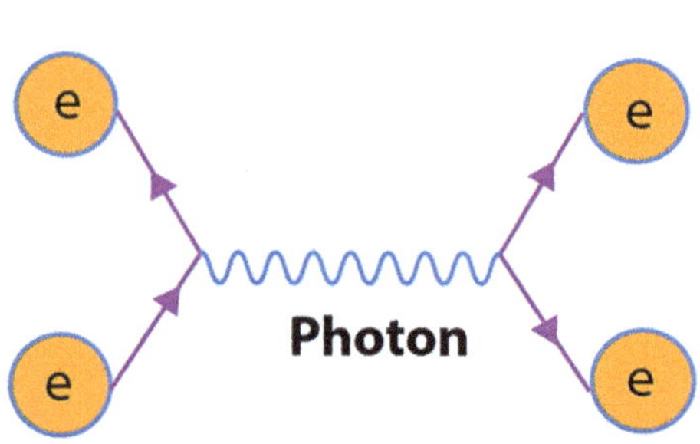

Figure 2.9. Electrons interact with each other through exchange of photons.

Forces between protons and neutrons are mediated through exchange of gauge bosons which results in transformation of quarks to up and down. For example, a down quark in a neutron can get transformed into an up quark through emission of a gauge boson called W^- boson (figure 2.10). Similarly, an electron is transformed into a neutrino by emission of W^+ boson (figure 2.11).

The quarks change their flavour from one form to another by emitting W bosons (figure 2.12). This aspect is called quantum flavour dynamics. The symmetry of transformation of quarks is defined by a mathematical formalism called SU(2) symmetry. Here, the quantities which are conserved are isospin and hypercharge which are physical quantities which distinguish different kinds of quarks.

The strong interactions between quarks are mediated by eight different kinds of bosons (gluons). Exchange of gluons can change the colour of quarks. Here the colour is the physical quantity which is conserved. Figure 2.13 illustrates the transformation of colour in quarks under exchange of a green-antiblue gluon.

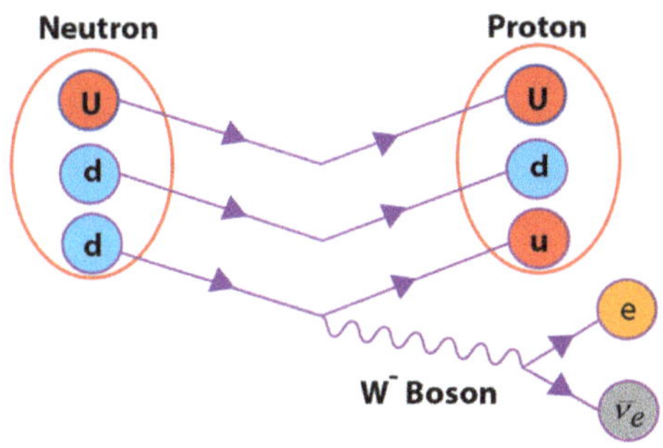

Figure 2.10. Transformation of neutron into proton.

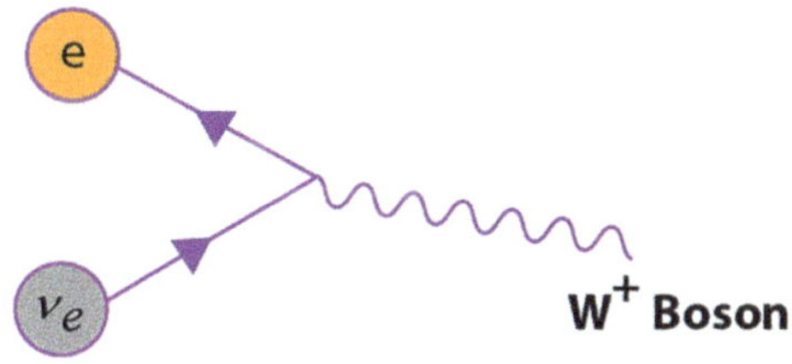

Figure 2.11. Transformation of electron into neutrino.

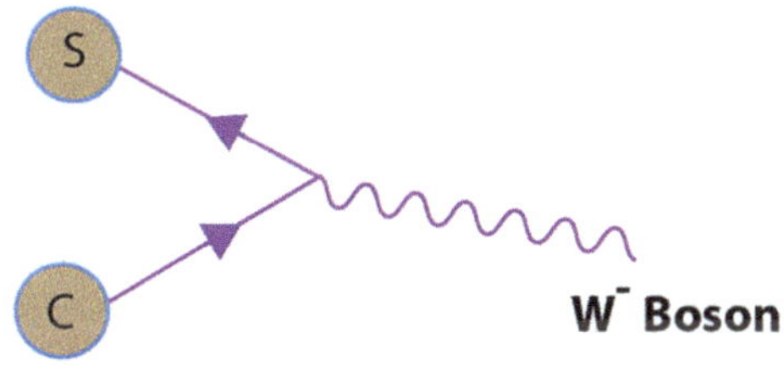

Figure 2.12. Transformation of flavour in quarks through emission of boson.

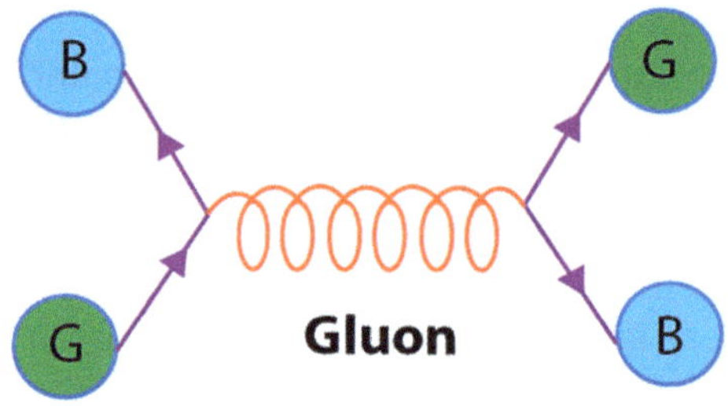

Figure 2.13. Interaction between quarks through gluons.

The U(1) symmetry group which describes electron–photon interaction is Abelian in nature and the gauge boson is the photon. The details of the process are described under a field widely referred to as quantum electrodynamics (QED). The gauge field in QED is basically the electromagnetic potential. The SU(2) group represents the rotational symmetry of a sphere. It comprises a set of 2×2 matrices having a unity determinant. They denote weak nuclear forces between fermions and bosons which are mediated via gauge bosons causing a change in the flavour of the quarks. This aspect is bundled in a subject called quantum flavour dynamics. In a similar manner, SU(3) represents interaction between strong forces between quarks which are mediated by gluons and they change the colour of the quarks. This aspect covering the transformation of colour of quarks is called quantum chromodynamics and it deals with conservation of colour in quarks. The entire set comprising all gauge transformation is called gauge group. All the elements of the group can be expressed using a function which establishes a mapping between the space–time variables and the Lie group. This function is special in the sense that its values and all the derivatives are the action of the gauge transformation at some specific points in space–time manifold. It is said to signify the symmetry of the gauge transformation. It was modified resulting in a non-Abelian gauge theory called the Standard Model whose symmetry group is called U(1) × SU(2) × SU(3). The total number of gauge bosons in this model are twelve, comprising eight gluons which determine the strong force defined by SU(3) group, three weak bosons responsible for the weak force described by SU(2) group and the photon which is responsible for force between charged particles which is expressed by SU(1) symmetry group.

The strong interactions in subatomic physics which hold together positively charged protons remained a mystery for long until Yang and Mills discovered a new model which is also called a non-Abelian gauge theory laying down the foundation of the Standard Model. These are based on non-commutative groups which describes a field which interacts with itself. It is worth mentioning that electromagnetic fields do not interact with themselves. Gravitational field is one such field which interacts with itself as the field has energy and energy is equivalent to mass and mass is a source of gravitational field. The nuclear forces also have the same self-interacting property.

A student called Ronald Shaw while working under Abdus Salam at Cambridge University also discovered the theoretical framework of non-Abelian guage theory during his doctoral work. However, the Yang–Mills model received a lot more popularity and recognition. The model generalised the gauge invariance of electromagnetism while placing special emphasis on quantization of the gauge field while focusing attention on the action of SU(2) symmetry group on subatomic particles like protons and neutrons. The ideas were refined by Salam, Weinberg and Glashow which resulted in a unification of electromagnetic forces along with weak forces. It led to the discovery of the concept of asymptotic freedom by Gross and Wilcek which described strong interactions and eventually led to a gauge theory which described strong interactions. This is now called quantum chromodynamics which encompasses action on the SU(3) group. It completed the key aspects of the Standard Model which successfully describes electromagnetic force, the strong and the weak force in subatomic physics. The Standard Model is a gauge theory with three symmetry groups called the SU(3) × SU(2) × SU(1).

An important aspect of gauge theory is to relate local symmetry to global symmetry. A gauge field is introduced and all the rates of change are discussed in terms of its covariant derivative. It represents the energy term. Gauge theory also finds a prominent place in describing gravity in the general theory of relativity. Here, the gauge field is called the Lanczos tensor. A quantum mechanical version of the theory of gravitation is based on gravitons which is the gauge boson responsible for gravitational potential. In general relativity, the transformation of a coordinate system results in invariance of a set of physical quantities associated with the dynamics of a system. This is called the principle of covariance, which allows choosing a general coordinate system to describe a certain dynamics. This is also called diffeomorphism. Gauge symmetries are analogous to this aspect in general relativity which allows invariance with a change in a certain variable. Attempts have been made to provide a new theory of gravity which is called the gauge theory of gravity which proposes to replace the concept of covariance with new gauge fields.

2.5 Conclusion

The concept of symmetry developed as a symbol of aestheticism and developed to express architectural forms which were implemented in great historical landmarks across all the ages. Towards the beginning of the modern era, its mathematical form was defined in terms of invariance of physical quantities which found great acceptance in describing the physics of a dynamic system. With time, the basic ideas of symmetries and associated conserved currents evolved from description of particle dynamics into complex forms while incorporating the dynamics of different kind of fields. The various forms of the Lagrangian subsume the subtle aspects of a system's dynamics and express the related conservation currents while offering a framework towards understanding the physics of different kinds of field interactions. The mathematical framework related to the Lagrangian of different kinds of fields has governed the physics of the 20th century.

References

[1] Stewart A 1978 'Polykleitos of Argos', One Hundred Greek Sculptors: Their Careers and Extant Works *J. Hellenic Stud.* **98** 122–31

[2] Galen 1981 *On the Doctrines of Hippocrates and Plato* (Berlin: Akademie-Verlag) 5.3.15

[3] Hon G and Goldstein B R 2005 From proportion to balance: the background to symmetry in science *Stud. Hist. Phil. Sci.* A **36** 1–21

[4] Vitruvius Pollio, Cesariano C and Krinsky C H 1969 *De Architectura* p 25 (Munchen: Wilhelm Fink)

[5] Pollio M V, Perrault C and Tardieu E 1859 *Les Dix Livres d'Architecture de Vitruve* (Paris: A Morel)

[6] https://commons.wikimedia.org/wiki/File3ADa_Vinci_Vitruve_Luc_Viatour.jpg

[7] Hon G and Goldstein B R 2005 Legendre's revolution (1794): the definition of symmetry in solid geometry *Arch. Hist. Exact Sci.* **59** 107–55

[8] Weyl H 1952 *Symmetry* (Princeton, NJ: Princeton University Press)

[9] Weyl H 1997 *The Classical Groups: Their Invariants and Representations* vol 1 (Princeton, NJ: Princeton University Press)

[10] Gross D J 1996 The role of symmetry in fundamental physics *Proc. Natl Acad. Sci.* **93** 14256–9

[11] Noether E and Noether E 1981 A Tribute to her Life and Work ed J W Brewer and M K Smith *Monographs and Textbooks in Pure and Applied Mathematics* 69 (New York: Marcel Dekker)

[12] Noether E 1918 Invariante variationsprobleme *Nachrichten von der Gesellschaft der Wissenschaften zu Göttingen, mathematisch-physikalische Klasse* **1918** 235–57

[13] Quigg C 2013 *Gauge Theories of the Strong, Weak, And Electromagnetic Interactions* (Princeton, NJ: Princeton University Press)

[14] Goldstein H 1965 *Classical Mechanics* (Reading, MA: Addison-Wesley)

[15] Adair R K 1987 *The Great Design: Particles, Fields and Creation* (New York: Oxford University Press)

[16] Sakurai J J 1964 *Invariance Principles and Elementary Particles* (Princeton, NJ: Princeton University Press)

[17] Flamm D and Schöberl F 1982 *Introduction to the Quark Model of Elementary Particles: Quantum numbers, gauge theories, and hadron spectroscopy* vol 1 (New York: Gordon & Breach)

Dhiraj Sinha and Gehan A J Amaratunga

Chapter 3

Spontaneous symmetry breaking

3.1 Symmetry breaking

Symmetry breaking can be defined as a process where a dynamic system undergoes a set of transformations from its initial stage of symmetry to a final stage of asymmetry, such that one or more physical quantities defining the properties of the system lose their corresponding states of invariance, leading to the asymmetric state. For example, a circle is a locus of points that are equidistant from the centre and have rotational invariance; if the circle is transformed into a line, the corresponding symmetry of the equidistant points is lost, leading to symmetry breaking. It is worth mentioning that in symmetry breaking, not all symmetries are lost. For example, during the transformation from a circle to a line, the initial symmetry of the equidistant points is lost, but the line has its own symmetry, defined in terms of the set of collinear points, which have translational symmetry (figure 3.1). Thus, symmetries are not lost in symmetry breaking, but are lowered from a state of higher order to a state of lower order.

There are two kinds of symmetry breaking in physical systems—explicit symmetry breaking and spontaneous symmetry breaking [1]. In explicit symmetry breaking, the symmetry is explicitly broken by introducing some changes into the structure or dynamics of a system, resulting in the loss of invariance for a set of physical quantities that define the invariance of the system. These changes are also reflected in the defining equations of the system. In spontaneous symmetry breaking, the dynamics of the system defined by its Lagrangian has symmetry, but the symmetry is lost for its lowest energy states, which correspond to broken symmetry [1]. To illustrate the phenomenon of spontaneous symmetry breaking, we consider the case of the symmetry of a cylindrical rod standing vertically. When axial force is applied on it from the top, the rod can become buckled to one of its low energy states under the given load, resulting in spontaneous symmetry breaking (figure 3.2). There are a number of possibilities in terms of the geometric directions along which such a

doi:10.1088/978-0-7503-5132-4ch3

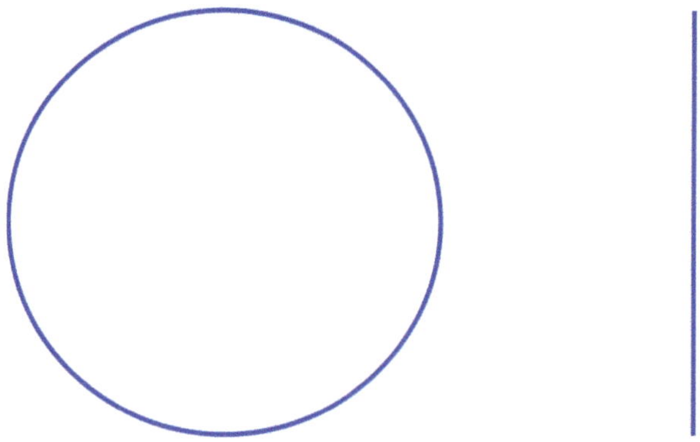

Figure 3.1. The symmetry of a circle is defined by the locus of points that are equidistant from the centre and have rotational invariance. The symmetry can be broken by shifting the points along its diameter in the form of a straight line, where the initial symmetry is lost, but the set of points are defined by translational invariance. Thus, in symmetry breaking, symmetries are not completely lost, but transformed into a state of lower order.

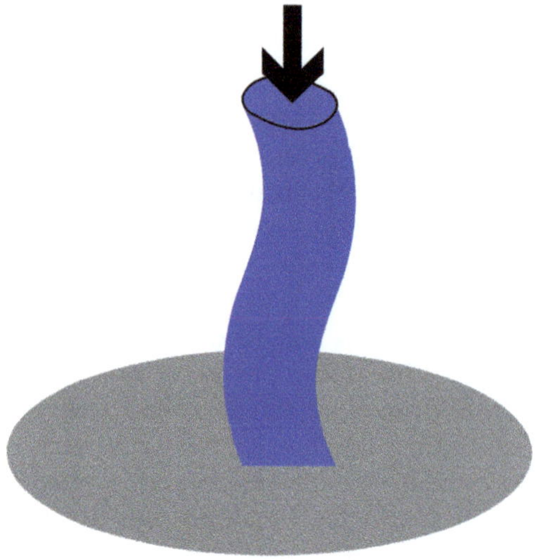

Figure 3.2. A vertically standing rod under an axial force is in a state of symmetry that is spontaneously broken after some time and the rod bends to one of its lower symmetry states.

buckling can happen. But the rod would spontaneously settle down to one of the possible final states under the given load while losing its initial symmetry. If the frictional forces are absent, the rod can be spun without any applied energy along its vertical axis. This spinning mode is an instance of the Nambu–Goldstone boson arising out of spontaneous symmetry breaking [2]. Another related example is that of a ball on top of a hemispherical dome with a trough lining the bottom of the structure. The system is symmetrical, but as the ball is released from rest it can reach one of the many possible low energy states, resulting in a state of spontaneously broken symmetry. Although the ball and the hemispherical dome along with its trough retain their distinct symmetry, the entire structure is asymmetrical.

The origin of spontaneous symmetry breaking was the theory of ferromagnetism proposed by Heisenberg in 1928, which postulates an infinite array of magnetic moments with spin interactions among the neighbours that result in a net preferential alignment [3]. The translational and rotational invariance vanishes below the Curie temperature and there is spin alignment in a certain specific direction. Below the Curie temperature, there are an infinite number of degenerate ground states where the spins are aligned in a certain direction. However, the Hamiltonian still has translational symmetry in the local region, it remains hidden. The same model is leveraged in the context of quantum field theory, where the ground state has broken symmetry but the system's Lagrangian is still symmetric.

3.2 Historical overview and early evolution

The initial work in the field of spontaneous symmetry breaking was performed by Nambu in the context of superconductors in 1960 [4]. Nambu's main argument was that the ground state of the electron wavefunction in a superconductor is associated with spontaneous symmetry breaking defined in terms of the loss of gauge invariance (also referred to as the breaking of U(1) symmetry), which generates massless excitations of electron–hole pairs (Cooper pairs) which propagate as modes that are now widely referred to as Nambu–Goldstone modes.

This aspect was particularly highlighted in his Nobel lecture [2]: 'One day before publication of the BCS paper, Bob Schrieffer, still a student, came to Chicago to give a seminar on the BCS theory in progress .... I was very much disturbed by the fact that their wave function did not conserve electron number. It did not make sense .... At the same time I was impressed by their boldness and tried to understand the problem.' He continued, 'The BCS theory assumed a condensate of charged pairs of electrons or holes, hence the medium was not gauge invariant. There were found intrinsically massless collective excitations of pairs [Nambu–Goldstone modes] that restored broken symmetries, and they turned into the plasmons by mixing with the Coulomb field.'

Under room temperature conditions, the Fermi surface in a material is stable, as it is not perturbed by external interactions. For example, when impurities are added to it, the Fermi surface does not undergo a change in its topology. When some structural changes are made in the system, it adopts a new stable topology. The Fermi surface holds translational and rotational symmetry with regard to changes that conserve the charges. The physics of the system undergoes some change at ultra-low temperatures, when Cooper pairing happens, as the initial symmetry of the fermions is broken, which results in the generation of bosons at the vacuum state. It could be argued that most of the interactions associated with the Fermi surface conserve the number of charge particles. But the formation of Cooper pairs leads to transmutation of particles which results in non-conservation of the number of fermions. Thus, the symmetry of the Fermi surface is broken due to instability of the Fermi surface resulting in in Cooper pairing. As the electrons at the Fermi surface having a symmetry expressed by the unitary U(1) matrix, disappear, some kind of energy gap is developed. The Cooper pairs are bosons which do not have a Fermi surface. As there are two kinds of bosons which are developed with opposite spins—the hole-like and the electron-like, the resulting symmetry is called doubled Z2 symmetry generated as a result of breaking of U(1) symmetry. In other words, it is assumed that in superconductors there is a field that breaks the gauge symmetry, leading to gauge invariance, which generates the Cooper pairs that are the corresponding Nambu–Goldstone modes.

An alternative way of looking at symmetry breaking of U(1) gauge symmetry in a superconductor is associated with the symmetry breaking of the electron wavefunction. In a normal metal, the phase of the wavefunction fluctuates in a random manner under scattering with each other and from phonons resulting in global symmetry. When the metal becomes a superconductor, the symmetry of the electron wavefunction is broken and there is a single phase existing throughout the metal [5] We can say that the U(1) symmetry is spontaneously broken. This aspect is evident in Josephson effect discovered by Brian Josephson, who noted that symmetry constraints imply that the absolute phase of electron wavefunction in an isolated superconductor cannot be observed. However, he discovered that if a weak link is provided between two superconductors, the phase difference could be measured [6]. This weak link is called Josephson link, which comprises a superconductor separated by an insulating substance which allows tunneling of the electrons. Under such a condition, the coupling energy is periodic in ϕ because the observable is invariant under substitution of $\varphi \rightarrow \varphi + 2\pi n$ where n is an integer. The Josephson current is proportional to the sine of the phase difference of the electron wavefunction across the junction and is expressed by, $I = I_m \cos \varphi$ where I_m is maximum current in the superconducting loop. The relationship between voltage U and phase in a Josephson link are $U = (h/4\pi e)\partial\varphi/\partial t$ where h is Planck's constant, e is charge of electron and t is time.

The concept of spontaneous symmetry breaking was applied in the context of high energy physics by Nambu who postulated that chiral symmetry breaking results in generation of pions; the lightest meson [7]. The ideas were further refined and developed by Goldstone, Salam and Weinberg. The fundamental physical aspect of the generalisation is that when a continuous symmetry of a dynamic system is spontaneously broken, the associated Noether currents are conserved, but the ground state of the system is not invariant and loses its symmetry [8, 9]. Under such a condition, massless particles appear. For every symmetry breaking one Nambu–Goldstone boson is generated, whose wavelength is associated with the fluctuation of the corresponding order parameter. This came to be called the Goldstone theorem, according to which, with every spontaneous symmetry breaking, massless bosons would be generated. A related aspect is that these bosons are excited by vacuum field. They are called Goldstone or Nambu–Goldstone bosons.

Besides superconductivity, spontaneous symmetry breaking covers a rather broad spectrum of physical phenomenon. In crystals, spontaneous breaking of translational and rotational symmetry causes the generation of waves called Goldstone phonons. As a liquid is solidified into a solid, its translational and rotational symmetry (also called Galilean symmetry) is spontaneously broken. The corresponding Goldstone modes are phonons propagating in longitudinal and transverse directions. As a ferromagnetic material is cooled below the Curie temperature, the spontaneous symmetry breaking results in a net magnetisation which in turn results in the generation of spin waves called magnons, which are the associated Goldstone modes [10]. In the Higgs mechanism of the standard model, particles are assumed to obtain mass due to the spontaneous breaking of local symmetry [11]. Thus, the phenomenon can explain phases of diverse kinds of materials like superconductors, magnets and crystals, along with their phase transitions. However, there are some phenomena that cannot be explained by it. The fractional quantum Hall effect, where the Hall voltage shows a fractional shift with magnetic field, is a notable exception.

The field of spontaneous symmetry breaking has been recognised with a number of Nobel prizes. The 2008 Nobel Prize in Physics was awarded for the study of symmetry breaking at sub-atomic dimensions. The first half of the prize went to Yoichiro Nambu of the University of Chicago for his work on spontaneous symmetry breaking in the context of chiral symmetry breaking, which is a strong interaction. The other half of the Nobel Prize was awarded to Makato Kobayashi and Toshihide Maskawa for their work on the explicit breaking of charge parity symmetry in weak interactions of the standard model, which predicts at least three families of quarks. In 2013, the Nobel Prize in Physics was jointly awarded to Peter W Higgs and Francois Englert for their work on the mechanism for particles acquiring mass. This is also widely regarded as the Higgs mechanism, which is a modification of the Goldstone theorem associated with spontaneous symmetry breaking. The fundamental work of Higgs was to show that under certain conditions of symmetry breaking, the Goldstone bosons will acquire a finite amount of mass. Further modifications of the idea were made by Englert and Brout. Other scientists, such as Kibble, Guralnik and Hagen, also made notable contributions, although they were not considered by the award committee. Higgs named his Nobel Lecture, 'Evading the Goldstone theorem' [12]. It was a notable tribute to Jeffrey Goldstone, who was excluded from the award committee but whose initial work in spontaneous symmetry breaking heralded a new chapter in theoretical and high-energy physics.

3.3 Symmetry breaking in particle physics

Spontaneous symmetry breaking is of the utmost importance in the field of particle physics. An important aspect of the dynamic equations of particle physics known as the field equations is characterised by a symmetry called gauge symmetry, which gives the freedom to choose the electric field and vector magnetic potential of an electrodynamic system, thus freeing it from the constraints of specific spatial and temporal coordinates. The field equations make symmetry predictions about the outcome of experiments. For example, they predict that the mass of two particles will be the same at the end of a certain experiment. When the equations are solved for specific cases, it appears that in one solution, one particle is heavier than the other, and in another solution, the opposite happens, while the difference in the mass remains the same. The symmetry inherent in the field equations is present in the entire spectrum of solutions, but it is absent in the distinct solutions. When a particular experiment is carried out, the system leads to one particular outcome that is recorded by the measuring instruments, which is an empirical instance of spontaneous symmetry breaking where the initial symmetry appears to be broken. The fact is that the initial symmetry remains hidden from the empirical observations of the experiment, which is widely referred to as broken symmetry.

The phenomenal observations of spontaneous symmetry breaking gradually found increasing acceptance in the field of particle physics. An important instance of spontaneous symmetry breaking in this domain is the chiral symmetry breaking associated with strong interactions in particle physics. The part of quantum field theory that describes strong interactions is called quantum chromodynamics and it describes the mechanism of the mass of subatomic particles like mesons and baryons. Chiral symmetry breaking results in the generation of Nambu–Goldstone bosons which are called pions [2, 7]. These pions, which are made of one quark and one anti-quark, are basically pseudo-Goldstone bosons, as they are a result of spontaneous as well as explicit symmetry breaking. There is a spontaneous symmetry breaking of chiral symmetry under strong interactions. Quark condensation plays a key role here and the symmetries are further broken by the mass associated with quarks. The resulting pions are not massless, but the value of their mass is lower than that of hadrons.

In 1965, Peter Higgs met Jeffrey Goldstone at CERN during a summer fellowship. He immediately became fascinated with the idea of spontaneous symmetry breaking [12]. He made some modifications to the initial ideas of spontaneous symmetry breaking propounded by Goldstone. Here, the symmetry breaking is such that the ground state of the symmetric system has a broken symmetry, while the dynamic equations represented by the Lagrangian are symmetric. In the context of gauge symmetry, unlike the theories postulated by Goldstone, where the vacuum based perturbations result in symmetry breaking that results in an unstable particle, here, Higgs particles are not unstable, but the unstable fields are generated by some bound states associated with the system. Fermionic condensates like quark condensates can also cause dynamic symmetry breaking.

The $SU(2) \times U(1)$ gauge symmetry is an integral part of subatomic physics which describes electroweak interactions. However, the inherent symmetry leads to a problem, as the standard model predicts a number of particles which have not been found through experiments. In addition to that, some of the key particles, such as the W and Z bosons, should remain massless. But experimental results indicate that these are massive particles. This implies that the $SU(2) \times U(1)$ gauge symmetry must show some deviations in its form. To overcome these problems, a new theory was developed, which is currently referred to as the Higgs mechanism.

The key aspect of the Higgs mechanism is the addition of a special field called the Higgs field to the mathematical structure of the standard model. In the standard model, the particles are massless at high temperatures with electroweak symmetry. The W and Z bosons have a polarisation component along the longitudinal direction. They correspond to Goldstone modes of spontaneous symmetry breaking of the electroweak symmetry ($SU(2) \times U(1)$). These are not observable. The gauge bosons are annihilated, which eventually gives a mass to the bosons. At certain temperatures, the symmetry is broken spontaneously as a result of interactions with the scalar field and the W and Z bosons become massive. The symmetry breaking causes the Higgs mechanism, which results in the bosons gaining a finite value of mass. In more specific terms, the electroweak symmetry is broken and as a result the W and Z bosons gain mass. It is also referred as electroweak symmetry breaking. A colourful way to express this is to say that the Goldstone bosons are eaten up by the gauge bosons, which eventually become massive. Other particles, such as leptons and quarks, can also acquire mass using the Higgs mechanism, although in a different way from the gauge bosons.

A non-relativistic effect related to the Higgs mechanism is seen in superconductivity, where the bosons develop a dynamic mass that can exclude flux from a superconductor, which is expressed through the Ginzburg–Landau theory. The extremely massive plasma mode in a superconductor is another instance of the Higgs boson [13]. Here, the U(1) gauge symmetry is spontaneously broken, resulting in the generation of Cooper pairs.

An important aspect of the standard model is that photons are responsible for electromagnetic forces, while the W and Z bosons mediate the weak forces. At energies above 100 GeV, the particles cannot be distinguished from each other. Weinberg and Salam have theorised that at low energies, the existing symmetry is broken, resulting in the emergence of photons and W and Z bosons.

The basic ideas associated with the Higgs mechanism were proposed by Anderson in 1962, based on the work of Nambu and Goldstone. A more refined theory was presented by three different groups in 1964, including Robert Brout and Francois Englert of the University of Brussels, Peter Higgs of the University of Edinburgh and Gerald Guralnik, C R Hagen and Tom Kibble of Imperial College, London. On 14 March 2013, CERN announced that the Higgs boson had been found at the Large Hadron Collider.

3.4 Condensed matter, superfluidity and Bose–Einstein condensate

The main ideas of spontaneous symmetry breaking evolved in the context of condensed matter physics, where understanding the phase changes of crystals and related materials using existing theories was proving to be difficult. For example, a material in a liquid state is characterised by symmetry as matter is uniformly distributed along all the directions. When it is solidified, the initial symmetry along the rotational and translational directions is lost as a crystal is formed. As the unit cell is repeated uniformly in three-dimensional space, the system has some kind of symmetry, which has a lower order than its initial state. This is an example of spontaneous symmetry breaking. Another interesting example can be cited from magnetism in a material at room temperature, where the magnetic poles are randomly aligned in all directions without any preferential orientation. The magnetic energy of the system does not vary with a change in position. So, there is translational as well as rotational symmetry of the magnetic field. When the magnetic material is cooled below the Curie temperature, there is spontaneous magnetisation, which results in the preferential orientation of the magnetic fields along certain directions. This leads to a condition where the Hamiltonian's value changes with position and orientation as the system has a net finite value of magnetism [10]. We can say that the initial symmetry has been spontaneously broken as the material has cooled. The concept has been used to explain a number of related phenomena, such as phase changes in liquid crystals, the generation of superfluidity and the propagation of magnetic waves called magnons, along with polarisation waves, in piezoelectric materials.

All the systems that exhibit spontaneous symmetry breaking possess a physical quantity that controls the initial symmetry. It is called the order parameter and it can be used to quantify the process of symmetry breaking. In the simplest terms, it is a variable that acquires a final value which is not invariant from its initial state of invariance. In magnetic materials, the value of local magnetisation is the order parameter. Order parameters show long wavelength fluctuations with every spontaneous symmetry breaking and do not need any energy for their excitation. These are widely referred to as Goldstone or Nambu–Goldstone bosons, as they resemble normal bosons and are massless. In crystals, the order parameter is the spacing between the crystals that results in a change in density of the crystal atoms and the Nambu–Goldstone bosons are the phonons. In the context of magnets, these modes are magnons or spin waves.

There are cases where spontaneous symmetry breaking can be prevented by related physical effects. For example, at non-zero temperatures in single-dimensional systems, the order is destroyed by the Nambu–Goldstone modes taking the system to its initial stage of symmetry. Similarly, fluctuations linked with quantum mechanical perturbations can prevent symmetry breaking in one-dimensional systems like nanowires. Symmetry breaking has also been found to be induced by interactions associated with the Yukawa potential. Similarly, the Coulomb potential can also break the symmetry of some systems. For a long time, it was assumed that spontaneous symmetry breaking could explain all the aspects of phase changes in matter. However, the discovery of the fractional quantum Hall effect changed things. Similarly, spin liquids are another related class of matter where these ideas cannot be applied. The development of an integrated framework for understanding these phenomena is one of the biggest challenges for condensed matter theory.

There are two other key areas where spontaneous symmetry breaking appears—Bose–Einstein condensates (BEC) and superfluidity. Bosons, which are particles with integral spin, can share the same quantum state. It was Albert Einstein who proposed that cooling bosons to ultra-low temperatures could result in the lowest quantum state. This is the BEC. Way back in 1938, it was postulated by Fritz London that a BEC could result in superfluidity and superconductivity [14]. A BEC was realised at the experimental level in 1995 by Eric Cornell and Carl Wiemann in Boulder while working at the NIST-JILA lab. They cooled rubidium atoms to 170 nK. Wolfgang Ketterle at MIT achieved a BEC using Sodium-23, working on some key physical properties like quantum mechanical interference [14]. A BEC has also been observed in the case of lithium atoms by Randal Hulet at Rice University. The BEC is an instance of spontaneous symmetry breaking, as bosons in a general state possess symmetry. They lack preferred direction and the phases are random, which indicates some kind of invariance and symmetry. When the gas is cooled to temperatures close to absolute zero, the atoms adopt a similar quantum mechanical state, having the same phase, which corresponds to spontaneous symmetry breaking. This happens because of the spontaneous breaking of the gauge symmetry.

Superfluidity is a physical state of matter in fluid form where the viscosity and surface tension disappear. The liquid is able to flow while defying gravitational force and dissipation. The phenomenon has mainly been observed in the context of liquid helium. The initial discovery was made by Pyotr Kapista and John Allen in 1937 [15]. In helium-4 each atom is actually a boson with an integer spin, but in helium-3 each atom is a fermion with fractional spin. Superfluidity is achieved in helium-4 at higher temperatures than in helium-3, where the fermions are transformed into bosons at low temperatures through spontaneous symmetry breaking. Superfluidity was discovered in helium-3 by Douglas Osheroff, David Lee and Robert Richardson in 1976, for which they received the Nobel Prize in 1996. In recent years, superfluidity has also been discovered in fermionic gases like lithium at temperatures of the order of 50 nK. The first result in the field was reported by the Ketterle group at MIT in 2005, who recorded the formation of quantum vortices and quantum solitons in such gases.

In superfluidity, the global phase symmetry is spontaneously broken. The wavefunction of the condensate has a unique phase at each point. The corresponding Lagrangian of the system's symmetry is defined by phase invariance, which has an associated conserved quantity that is the particle number. The particle number is also referred to as the probability current by many authors. This particle conservation, which is reflected in phase invariance of the associated wavefunction, is also called U(1) global symmetry. In the superfluid state, the initial phase symmetry of the wavefunction is spontaneously broken.

3.5 Spontaneously broken global symmetry

There are some general features of spontaneous symmetry breaking which apply to all the systems. The basic aspect is that the system is such that its initial state can undergo spontaneous changes without any use of explicit force and in the process of the change, several equally probable final states are possible. So, in the initial phase, the system has a symmetry associated with regard to the final states which it will adopt. When the system is left on its own, it will make a transition to one of the final possible states which is asymmetric. As each of the final states is equally probable, the system is said to have hidden symmetry even when the symmetry is spontaneously broken. In mathematical terms, the possible outcomes of final transition from its initial state form a symmetric group. However, one of the members of the group can acquire a distinct value at the end of the transition which cannot be predicted at a theoretical level.

The symmetry group could be continuous or discrete. For example, a circle has continuous symmetry with regard to rotational changes, but an equilateral triangle has discrete symmetries as its shape remains invariant only if it is rotated by 60 degrees. In the case of crystals, a regular pattern is repeated in space only after a certain interval and the symmetries associated with unit cells which are repeated in three-dimensional space form a discrete group. An interesting feature is that in the context of a one-dimensional quantum-mechanical system, symmetry breaking can involve only discrete states.

The fundamental physical quantity which is investigated in the studies on symmetry breaking is the boson field. It is a complex scalar field denoted by φ and is not equal to its conjugate, i.e. $\varphi \neq \varphi^*$ [16]. It can be decomposed into its real and imaginary parts,

$$\varphi = \frac{1}{\sqrt{2}}\left(\varphi_1 + i\varphi_2\right) \tag{3.1}$$

Here, φ_1 and φ_2 are real valued functions. The Lagrangian density can be written as,

$$£ = \frac{\partial\varphi^*}{\partial t}\frac{\partial\varphi}{\partial t} - \frac{\partial\varphi^*}{\partial x}\frac{\partial\varphi}{\partial x} - m^2\varphi^*\varphi \tag{3.2}$$

where x denotes position, t is time and m is mass of the particle. It can be written as,

$$£ = \frac{\partial\varphi^*}{\partial t}\frac{\partial\varphi}{\partial t} - \nabla\varphi^*.\,\nabla\varphi - m^2\varphi^*\varphi = \partial^\mu\varphi^*\partial_\mu\varphi - m^2\varphi^*\varphi \tag{3.3}$$

Here μ is an index which spans the dimensions of the given coordinate systems of space and time and ∂_μ is a shorthand expression for $\partial/\partial\varphi^\mu$. Equation (3.3) is called Klein–Gordon Lagrangian density for the complex field [16].

If a quartic term $\lambda(\varphi^*\varphi)^2/2$ which preserves the global phase symmetry is added to the Klein–Gordon Lagrangian density we get,

$$£ = \partial^\mu\varphi^*\partial_\mu\varphi - m^2\varphi^*\varphi - \frac{1}{2}\lambda(\varphi^*\varphi)^2 \tag{3.4}$$

Here, λ is a dimensionless quantity and has a positive value. The potential associated with the Lagrangian is,

$$U = m^2\varphi^*\varphi + \frac{1}{2}\lambda(\varphi^*\varphi)^2 \tag{3.5}$$

The Hamiltonian corresponding to the Lagrangian of equation (3.5) is bounded from below for a positive value of λ. It can be represented by the schematic representation of figure 3.3 which has a mass at the centre of a potential well.

However, when $\lambda < 0$, the minimum energy state corresponds to a value of $\varphi = \varphi_0$ where the potential is associated with the following condition,

$$\varphi_0^*\varphi_0 = \frac{m^2}{\lambda} \tag{3.6}$$

Equation (3.6), represents a circle in the complex φ plane where,

$$\varphi_0 = \frac{m}{\sqrt{\lambda}}e^{i\theta} \tag{3.7}$$

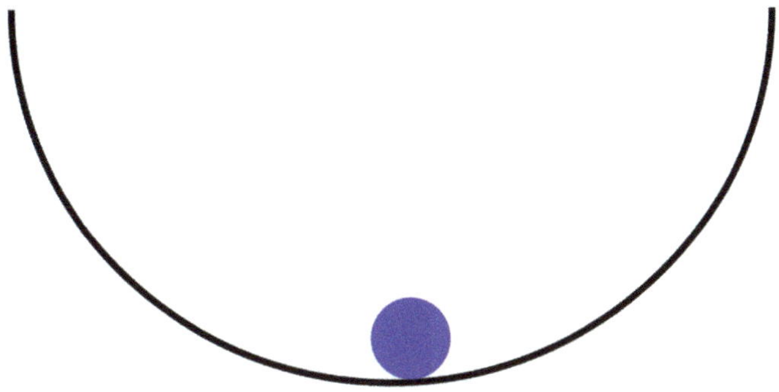

Figure 3.3. A mass at the bottom of a potential well indicating a symmetric dynamic system.

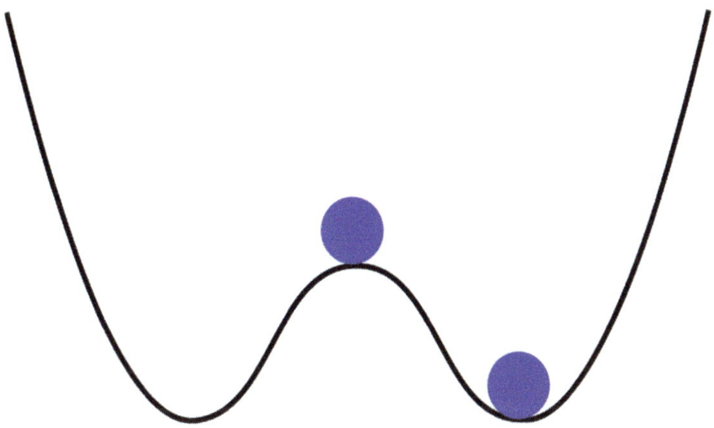

Figure 3.4. Front view of the base of a wineglass shaped potential. The ball kept at the centre of the potential well can roll to the trough breaking the symmetry of the system. The ball can roll spontaneously along the trough which corresponds to Goldstone modes.

The system has an infinite number of ground states having the same energy for different values of θ which indicates the phase of the scalar field. It corresponds to a physical situation where the system was in a state of symmetry initially as represented by the ball at the peak point of the bottom of a wineglass (figure 3.4) which drops to the trough spontaneously and rolls along its sides without any additional energy. The dynamics of the system in the initial phase is symmetrical with regard to rotations along the vertical axis. However, the final minimum energy state does not have a symmetry, as is represented in figure 3.5. Thus, the initial symmetry is spontaneously broken. The movement of the particle in the potential well does not need any energy and it corresponds to propagation of low energy Goldstone modes.

It is important to fix the value of θ and study the dynamics under variation of the field φ in order to evaluate the role of quartic potential in symmetry breaking. When $\lambda \to \infty$, the condition of Goldstone–Sombreao potential occurs and the Lagrangian adopts the least value. The action integral is symmetrical with U(1) transformation

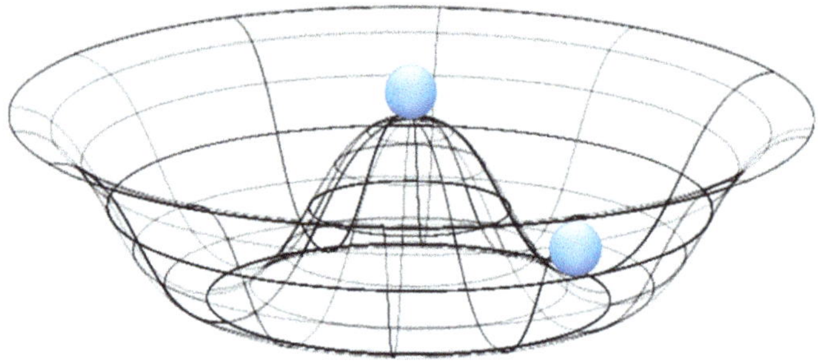

Figure 3.5. A three-dimensional representation of the potential energy landscape of figure 3.4.

which involves a change of phase. Here, θ corresponds to exciting the Nambu–Goldstone boson and U(1) symmetry transformation results in a phase shift. It does not preserve the ground state $|0\rangle$. In other words, the transformation does not annihilate the ground state. Here the vacuum has a degeneracy and is not invariant, which is the condition of spontaneous symmetry breaking.

Defining, $\varphi_0 = m/\sqrt{\lambda}$ and and substituting $\varphi = \varphi_0 + \chi$, to represent the impact of a dynamical field χ, we can write the potential energy as,

$$U = U(\varphi_0) + \frac{1}{2}\lambda\left[\varphi_0(\chi^* + \chi) + \chi^*\chi\right]^2 \tag{3.8}$$

While using, $\lambda\varphi_0^2 = m^2$ and using the following relation,

$$\chi = \frac{1}{\sqrt{2}}\left(\chi_1 + i\chi_2\right) \tag{3.9}$$

we get,

$$U = U(\varphi_0) + m^2\chi_1^2 + O\left(\chi^3\right) \tag{3.10}$$

where, $O(\chi^3)$ is a function representing the higher order terms. When $\varphi = 0$, the system's state is at a vacuum state which is unstable and possesses U(1) symmetry. The system adopts a specific value of $\varphi = \varphi_0$, the symmetry is lost or in other words, spontaneously broken. The quadratic term in equation (3.10) can be compared to the Klein–Gordon equation of motion, $\partial^\mu\partial_\mu\varphi + m^2\varphi = 0$ which leads to the dispersion relationship, $\omega = \sqrt{k^2 + 2m^2}$ for the field χ_1 where k is the wave number. It implies that following second quantization, the quanta of the field will have particles with mass equal to $\sqrt{2m}$ assuming natural units. For the field, χ_2, there is no quartic term and it corresponds to the existence of a massless boson called Goldstone Boson with the dispersion relationship, $\omega = k$. It propagates along the path of the minimum potential surface at the bottom of the potential energy surface. This is an instance of Goldstone's theorem which states that for every spontaneously broken global continuous symmetry, there is a field with massless quanta which is called a Goldstone field.

3.6 Higgs mechanism

The Klein–Gordon Lagrangian (equation 3.3) is invariant under a global phase change defined by,

$$\phi \rightarrow e^{-i\varepsilon}\phi \approx \phi - i\varepsilon\phi \tag{3.11}$$

and

$$\phi \rightarrow e^{i\varepsilon}\phi \approx \phi + i\varepsilon\phi \tag{3.12}$$

Here, global change in phase implies that the phase change ε is the same at all the points along the dimensions of space and time. In a related case, when we change the phase such that it is dependent on space and time coordinates, x^{μ}, we can write phase as, $\phi \rightarrow e^{-i\varepsilon(x)}\phi$. A transformation of this kind which depends on position is called gauge transformation. It implies that the symmetry parameter can be chosen to vary from place to place. Here, $\varepsilon(x)$ is not a field but a generalized symmetry parameter which is allowed to vary from place to place. As we vary the phase along the spatial dimension, we introduce a gradient which was not present before. As all the interactions of matter are associated with gauge symmetry, we want to cancel out the phase variations in order to maintain overall gauge symmetry. This leads to the following,

$$\partial^{\mu}\phi \rightarrow e^{-i\varepsilon(x)}\left[(\partial^{\mu}\phi) - i(\partial^{\mu}\varepsilon)\phi\right] \tag{3.13}$$

Hence, the free field Klein–Gordon Lagrangian can be written as,

$$\pounds_{KG} = \left[\partial_{\mu}\varphi^{*} + i(\partial^{\mu}\varepsilon)\varphi^{*}\right]\left[\partial_{\mu}\varphi - i(\partial^{\mu}\varepsilon)\varphi\right] - m^{2}\varphi^{*}\varphi \tag{3.14}$$

$$= \pounds - i\left(\partial_{\mu}\varepsilon\right)\left[(\partial^{\mu}\varphi^{*})\varphi - \varphi^{*}(\partial^{\mu}\varphi)\right] + \left(\partial_{\mu}\varepsilon\right)(\partial^{\mu}\varepsilon)\varphi^{*}\varphi \tag{3.15}$$

If we substitute, $A_{\mu} = A_{\mu} + \partial_{\mu}\varepsilon/e$ and $\partial_{\mu} \rightarrow D_{\mu} = \partial_{\mu} + ieA_{\mu}$ where D_{μ} is covariant derivative and A_{μ} is a four vector potential, the kinetic energy terms in the Lagrangian are replaced by $D_{\mu}\varphi D^{\mu}\varphi^{*}$ and the invariance is restored, or,

$$D_{\mu}\varphi \rightarrow \left[\partial_{\mu} + ieA_{\mu} + i\left(\partial_{\mu}\varepsilon\right)e^{-i\varepsilon}\phi\right] = e^{-i\varepsilon}D_{\mu}\phi \tag{3.16}$$

This leads to,

$$\pounds_{KG} = \left(D_{\mu}\varphi\right)^{*}D_{\mu}\varphi - m^{2}\varphi^{*}\varphi \tag{3.17}$$

In other words, the Lagrangian remains unchanged. Thus, the theory remains invariant only when we introduce the electromagnetic field. The final Lagrangian for Klein–Gordon equation can be written as,

$$\pounds_{KG} = \partial^{\mu}\varphi^{*}\partial_{\mu}\varphi - m^{2}\varphi^{*}\varphi + ieA_{\mu}[(\partial^{\mu}\varphi^{*})\varphi - \varphi^{*}(\partial^{\mu}\varphi)] + e^{2}A_{\mu}A^{\mu}\varphi^{*}\varphi \tag{3.18}$$

The third term is the interaction term while the fourth term, i.e. the quartic term is electromagnetic potential which preserves the gauge invariance. We can construct a gauge invariant dynamics of the electromagnetic field coupled to a field whose quanta are charged particles. It can be written as

$$\pounds_{\mathbf{KG}} = -\frac{1}{4}F_{\mu\nu}F^{\mu\nu} + \left(D_{\mu}\varphi\right)^{*}(D^{\mu}\varphi) - m^{2}\varphi^{*}\varphi \tag{3.19}$$

where, $F_{\mu\nu}$ is the electromagnetic vector potential and $D_{\mu} = \partial_{\mu} + ieA_{\mu}$. Substituting the potential term, we get,

$$\pounds_{\mathbf{KG}} = -\frac{1}{4}F_{\mu\nu}F^{\mu\nu} + \left(D_{\mu}\varphi\right)^{*}(D^{\mu}\varphi) - V(\varphi) \tag{3.20}$$

For minimum energy state, we have,

$$\left(ie\varphi_0 A_{\mu}\right)^{*}\left(ie\varphi_0 A^{\mu}\right) = \frac{\mathrm{e}^2 m^2}{\lambda}A_{\mu}A^{\mu} \tag{3.21}$$

which leads to the equation of motion,

$$\partial_{\nu}\partial^{\nu}A_{\mu} + 2\frac{\mathrm{e}^2 m^2}{\lambda}A_{\mu} = 0 \tag{3.22}$$

and the following dispersion relationship, $\omega = \sqrt{k^2 + 2\mathrm{e}^2 m^2/\lambda}$ where the quanta has a non-zero mass of $\omega = \mathrm{e}m\sqrt{2/\lambda}$.

The dynamical part of the scalar field, when the local phase symmetry is spontaneously broken is expressed as,

$$\varphi = \varphi_0 + \chi = \phi_0 + \frac{1}{\sqrt{2}}\left(\chi_1 + i\chi_2\right) \tag{3.23}$$

When, χ is non-zero, the covariant derivative of φ takes the following form,

$$D_{\mu}\phi = \frac{1}{\sqrt{2}}\left(\partial_{\mu}\chi_1 + i\partial_{\mu}\chi_2\right) + ie\phi_0 A_{\mu} + \cdots. \tag{3.24}$$

The terms related to χ_2 get removed by the following gauge transform,

$$A_{\mu} \rightarrow A_{\mu} - \frac{1}{\sqrt{2}\,\mathrm{e}\phi_0}\partial_{\mu}\chi_2 \tag{3.25}$$

It leaves the field strength tensor unchanged and we have the following Lagrangian density,

$$\pounds_{\mathbf{KG}} = -\frac{1}{4}F_{\mu\nu}F^{\mu\nu} + \mathrm{e}^2\phi_0^2 A_{\mu}A^{\mu} + \frac{1}{2}\left(\partial_{\mu}\chi_1\right)\left(\partial^{\mu}\chi_1\right) - V(\phi) \tag{3.26}$$

which leads to,

$$\pounds_{\mathbf{KG}} = -\frac{1}{4}F_{\mu\nu}F^{\mu\nu} + \frac{e^2m^2}{\lambda}A_\mu A^\mu + \frac{1}{2}(\partial_\mu\chi_1)(\partial^\mu\chi_1) - m^2\chi_1^2 - V(\phi_0) \tag{3.27}$$

Finally, we have an expression where, χ_2 gets completely absorbed. To put it in a colourful manner, the Goldstone field has been eaten up by the vector field leading to acquisition of mass. This is the central aspect of the Higgs mechanism. When a continuous symmetry is spontaneously broken, it leads to Goldstone boson which is eaten by gauge boson to give mass to the field. All the observables in the laboratory get mass under SSB. Those which get mass without SSB are not observed and they are extremely heavy. Higgs field is the radial displacement away from the equilibrium, angular displacement corresponds to excitation of the Goldstone mode.

Mass is oscillation of a field when it is homogenous and has infinite wavelength. If wavelength of a particular field is infinite, any displacement leads to oscillation of the entire field. It also corresponds to particles at rest as they have mass. In other words, mass corresponds to oscillation of a field when it is homogenous and has infinite wavelength. A phonon has no mass as the shift in atoms is not homogenous. Plasmons have mass as it corresponds to a homogenous shift of charges.

When the angular frequency, ω is a linear function of wavenumber, k, there is no dispersion e.g. unity slope of $\omega - k$. It if is not linear, there is dispersion. A massless excitation is the one with a linear dispersion relation. A massive excitation has a dispersion relation that does not become zero as $k = 0$, but has a curvature. Mass is energy associated with zero momentum. Zero momentum means the field does not vary, i.e. the field is homogenous. Mass is also connected with the curvature of the potential energy surface at the equilibrium point. As soon as the field is shifted away from the equilibrium point, it starts to oscillate. If $\omega \neq 0$ at $k = 0$, the value of ω at that value of k is the mass. The only way, $\omega = 0$ is when the potential energy is flat. When the field is shifted, there is no oscillation, but there is some energy, we call it mass. If the field is excited, there is oscillation.

The mechanism of generation of mass through annihilation of Goldstone modes is also called spontaneously broken gauge symmetry. This is because the calculation involving quantization involves choice of a particular gauge which breaks the existing gauge symmetry. But the overall theoretical structure retains the global phase symmetry which is locally broken.

References

[1] Brading K and Castellani E 2003 *Symmetries in Physics: Philosophical Reflections* (Cambridge: Cambridge University Press)

[2] Nambu Y 2009 Nobel Lecture: spontaneous symmetry breaking in particle physics. A case of cross fertilization *Rev. Mod. Phys.* **81** 1015

[3] Heisenberg W 1928 *Zeit. für Phys.* **49** 619–36

[4] Nambu Y 1960 *Phys. Rev.* **117** 648

[5] Josephson B D 1962 Possible new effects in superconductive tunnelling *Phys. Lett.* **1** 251–53

[6] Josephson B D 1974 The discovery of tunnelling supercurrents *Rev. Mod.* **46** 251

[7] Nambu Y 1995 A 'superconductor' model of elementary particles and its consequencies (Talk given at a conference at Purdue (1960)) reprinted in *Broken Symmetries, Selected Papers* ed T Eguchi and K Nishijima (Singapore: World Scientific)
[8] Goldstone J 1961 *Nuovo Cimento* **19** 154
[9] Goldstone J, Salam A and Weinberg S 1962 *Phys. Rev.* **127** 965
[10] Wagner H 1966 *Z. Phys.* **195** 273
[11] Higgs P 1966 *Phys. Rev.* **145** 1156
[12] Higgs P W 2014 Nobel lecture: evading the Goldstone theorem *Rev. Mod. Phys.* **86** 851
[13] Anderson P 1963 *Phys. Rev.* **130** 439
[14] Pitaevskii L P and Stringari S 2003 *Bose–Einstein Condensation* (no. 116) (Oxford: Oxford University Press)
[15] Tilley D R and Tilley J 1990 *Superfluidity and Superconductivity* (Boca Raton, FL: CRC)
[16] Peskin M and Schroeder D 1995 *An Introduction to Quantum Field Theory* (New York: Westview Press)

IOP Publishing

Dhiraj Sinha and Gehan A J Amaratunga

Chapter 4

Explicit symmetry breaking and electromagnetic radiation

4.1 Explicit symmetry breaking of electrodynamic systems

In explicit symmetry breaking (ESB), the symmetry of a system is broken in an explicit manner by a process that transforms a set of variables defining the symmetry or invariance of the system. Unlike spontaneous symmetry breaking, where the system adopts one of the many possible ground or vacuum states in a spontaneous manner, in ESB there is only one predefined state to which the system is transformed from its initial stage of symmetry [1]. ESB in a dynamic system can be illustrated by using the example of a vertically standing rod subjected to a horizontal force (figure 4.1(a)). Under the impact of an external force, the rod can be bent explicitly in a certain direction, leading to a state of broken symmetry.

A common aspect of explicit symmetry breaking in dynamic systems is that the dynamic equations and the associated Lagrangian of the system lose their invariance as a consequence of some terms that break the equation's symmetry [1]. The concept holds some special importance in the context of electromagnetism, where the electric lines of field hold some special symmetries that can be transformed or broken under specific conditions of excitation of the corresponding electrodynamic system. For example, it is well known that the electric field lines of force of isolated point charges hold rotational symmetry (figure 4.2). This is neatly expressed using Gauss's law $\nabla \cdot E = Q/\varepsilon_0$, where E is electric field, Q is the total charge and ε_0 is permittivity. Figure 4.2 shows the electric lines of field of an isolated point charge which radially emerge from the centre in a symmetric manner. The electric potential associated with the charge has rotational symmetry as movement along the radial direction results in the same values of the corresponding electric potential. When the charge is accelerated temporarily, the pattern of electric lines of force changes. This is schematically represented in figure 4.3, which illustrates a broken symmetry of the electric lines of

doi:10.1088/978-0-7503-5132-4ch4

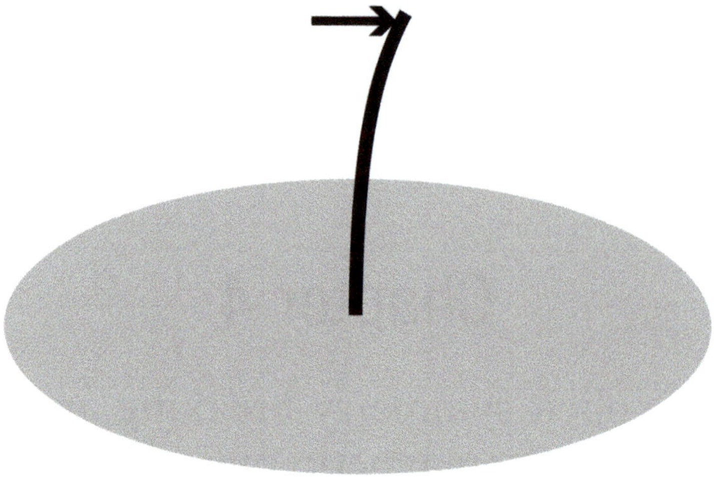

Figure 4.1. A rod in a vertical position has a symmetry that can be explicitly broken by an external force. Here, the final state is determined by the nature of the applied force.

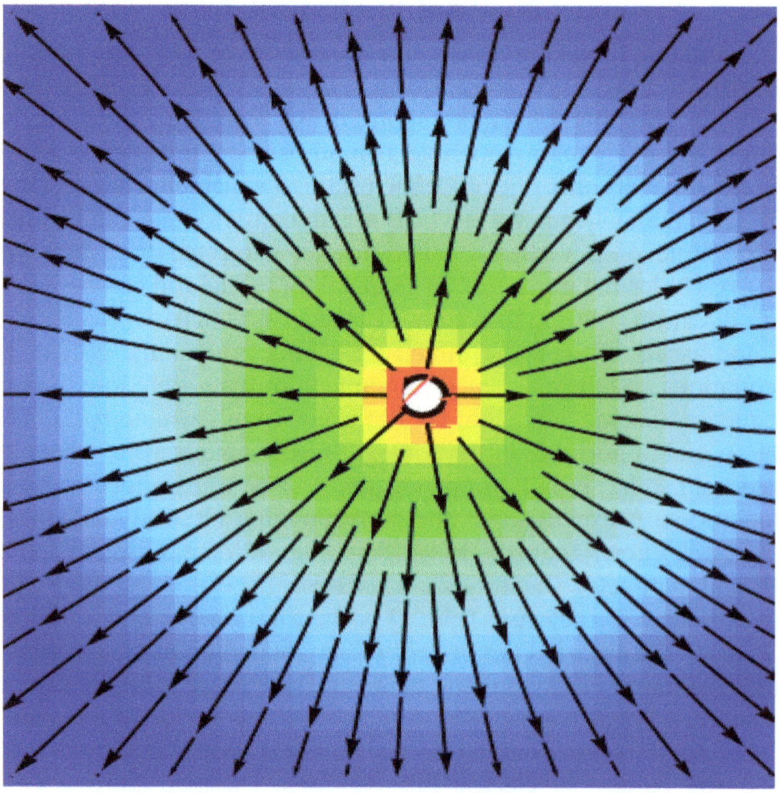

Figure 4.2. Symmetric pattern of electric lines of force from the centre of a positive charge.

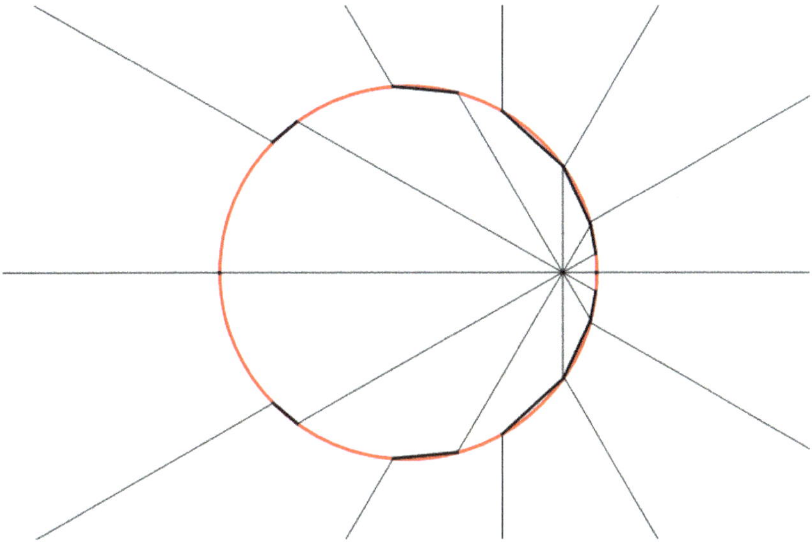

Figure 4.3. Pattern of electric lines of field around a temporarily accelerated charge.

Figure 4.4. Rotation of electric lines of field around a periodically accelerated charge.

field that results in variation of the corresponding electric potential. This aspect was first observed by Sir J J Thomson, who discovered the electron in 1896.

As the charge is periodically accelerated, the translational symmetry of the electric field is explicitly broken as a result of its rotation within a localised region of space and time [2]. This is illustrated in figure 4.4, where the rotating electric field lines are emerging from the centre of a periodically accelerated charged particle. This results in

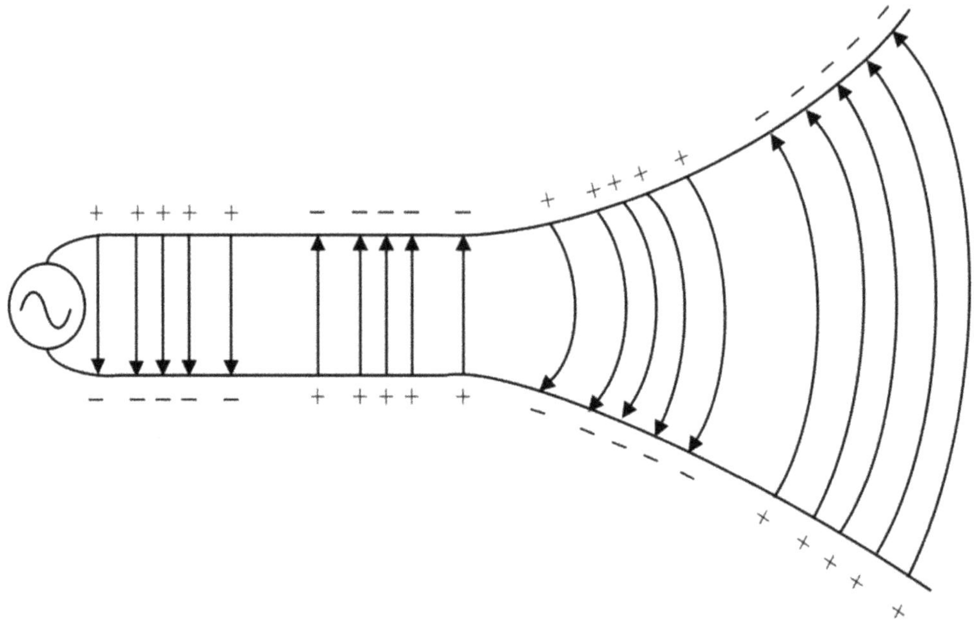

Figure 4.5. The electric field lines are straight in a parallel transmission line, indicating translational symmetry that is broken along the flared end in an explicit manner due to the geometric asymmetry of the system.

the generation of a time-varying magnetic field and an electromagnetic radiation in free space. The point form of Maxwell's equations, $\nabla \cdot \mathbf{E} = 0$ and $\nabla \times \mathbf{E} = -\partial\mathbf{B}/\partial t$, where **B** is magnetic flux density that is dependent on time t, describes the rotation of electric field and its association with time-varying magnetic field.

The geometric orientation of the electric and magnetic field lines of an electrodynamic system are governed by its geometric structure and the mode of excitation, which govern the symmetry or symmetry breaking of the field lines. This can be illustrated using a parallel two-wire transmission line with a time-varying voltage excitation fed along one of its terminals. The source voltage exciting the wire induces electrical charges of opposite polarities along the wire (figure 4.5). The electric lines of field associated with the opposite charges possess translational symmetry. Under time-varying electric excitation, there is acceleration of charge along the direction parallel to the transmission line, but as a consequence of the electrical constraints imposed by the symmetry of the transmission line, the translational symmetry of the electric field lines is maintained in the space between the transmission lines. The associated radiation fields are cancelled out and there is no emission of electromagnetic waves from the medium [3]. Such a condition is quite common in several instances of electrodynamic systems. For example, it is well known that a parallel plate or cylindrical capacitors do not emit or receive electromagnetic waves, despite excitation fed to them by a time-varying voltage source.

As the ends of a transmission line are opened up towards the ends (figure 4.5), the translational symmetry of the electric field existing within the inner space of the transmission line is explicitly broken. The rotational pattern of the electric lines of field (which are associated with the emission of electromagnetic radiation) is shown in figure 4.4.

4.2 Electromagnetic radiation under non-conserved Noether current

The relationship between symmetry and conserved quantities is outlined well in Noether's theorem, which says that for every symmetry existing in nature, some physical quantity is conserved [4]. The theorem can be applied easily in the context of static point charges, where the symmetry of the electric lines of force emerging radially from the centre is associated with conservation of charge in the localised region of space around the charge. The potential variation of the electric field lines around a static point charge has rotational symmetry or invariance. This is directly reflected in the angular invariance of the system's Lagrangian and its rotational symmetry. The potential energy linked to a test charge q at a distance r located in the field of point charge Q is given by, $U = Qq/(4\pi\varepsilon_0 r^2)$. The Lagrangian is expressed by, $\mathcal{L} = -U$. The temporal variance of the Lagrangian expressed by $\mathrm{d}\mathcal{L}/\mathrm{d}t$ is equal to zero and the same is the case with its angular variance expressed by $\mathrm{d}\mathcal{L}/\mathrm{d}\theta$, where θ denotes the angular displacement.

The relationship between symmetries in electric lines of field and conservation of charges can be extended to a parallel transmission line system, where the net symmetry of the electric field is associated with net conservation of charges within the localised region of space of the transmission line. In the context of a parallel two-wire transmission line excited by a voltage source at one end with the other ends floating in free space, despite electron acceleration along the transmission lines, the translational symmetry of the electric field is maintained at all times. For example, the electric lines of field emerging from the positive charges from the upper line end at the corresponding negative charges on the lower line and vice versa. This symmetry is associated with conservation of current within a localised region of space and time around the transmission line structure. For every element of charge which flows to the right end, there is a corresponding element of charge flowing to the left, resulting in zero net loss. If we neglect the resistive losses, the flow of charges can continue for an infinite time. The Lagrangian of the system also possesses translational symmetry. The magnetic field at a perpendicular distance r from the top wire is given by, $\mathbf{B}_1 = \mu_0 I_1/2\pi r$, where I_1 is the current flowing in it and μ_0 is the permeability of free space. For the lower wire with a current, I_2 the corresponding magnetic field is $\mathbf{B}_2 = \mu_0 I_2/2\pi r$. It is assumed that the wires are spaced very close to each other. Hence, for a distance $r \gg d$, the spacing between the wires, the total value of magnetic field is given by, $\mathbf{B} = \mathbf{B}_1 - \mathbf{B}_2 = \mu_0(I_2 - I_1)/2\pi r$. Here, the negative sign associated with I_1 indicates the phase difference of π radians between the current flowing in the two wires. Assuming the values of the current to be equal, the net magnetic field is zero. Assuming σ_1 as the charge density on the upper wire and σ_2 as the charge density on the lower wire, the net electric field at a distance r outside the wires is $\mathbf{E} = (\mathbf{E}_2 - \mathbf{E}_1) = (\sigma_2 - \sigma_1)/\varepsilon_0$, where ε_0 is the permittivity of free space. Its value is zero as the charge densities on the wires are similar in magnitude but opposite in sign. Thus, the net value of the Lagrangian of the electromagnetic field outside the wires is expressed by,

$$\mathcal{L} = \frac{1}{2}L(I_1 - I_2)^2 - \frac{1}{2C}(Q_1 - Q_2)^2. \tag{4.1}$$

Here, L is the inductance and C is the capacitance of the transmission line system. The net value of the Lagrangian is zero. The Lagrangian density of the electromagnetic field density outside the wires can be expressed as

$$\mathcal{L}_d = \frac{\mathbf{B}^2}{2\mu_0} - \frac{1}{2}\varepsilon_0\mathbf{E}^2. \tag{4.2}$$

Its value is zero as the values of the electric and magnetic fields are zero just outside the wire. This also implies the temporal and spatial invariance of the corresponding Lagrangian density. As far as the Lagrangian density of the electromagnetic field in the space within the transmission line is concerned, it has a net value expressed by,

$$\mathcal{L}_{\text{din}} = -\frac{1}{2}\varepsilon_0\boldsymbol{E}_{\text{in}}^{\;2}, \tag{4.3}$$

where E_{in} is the electric field within the transmission lines. The electric field has symmetries, which is reflected in its translational as well as temporal invariance, so the associated Lagrangian density remains invariant along the corresponding dimensions.

Symmetries are associated with conserved quantities, which implies that broken symmetries must be associated with the loss of the conserved values of some physical quantities within a localised region of space and time where the symmetry is broken. Thus, when the symmetry of the electric field of an electrodynamic system under time-varying excitation is broken, it results in electromagnetic radiation. For example, in the case of a parallel two-wire transmission line, the translational symmetry of the electric lines of field is broken by opening the wires towards one of its terminals (figures 4.5 and 4.6) and they form a curved shape towards the flared end of the transmission line, which is associated with electromagnetic radiation. In an electromagnetic wave propagating in free space, the electric lines of field curl around a central point (figure 4.4).

In electrodynamic systems which radiate in free space, current is not conserved within the temporal and spatial domain of the system, as it is lost during the process of the transformation of current into electromagnetic waves. The corresponding electric field tends to lose its translational as well as temporal symmetries, which is also reflected in the non-invariance of the values of its Lagrangian along the corresponding dimensions within the localised region of space and time. The non-conserved Noether current emerging from the localised spatial and temporal domain of the radiation system is transformed into electromagnetic radiation as a consequence of the broken symmetry of the electromagnetic field within the system. We can sum up by arguing

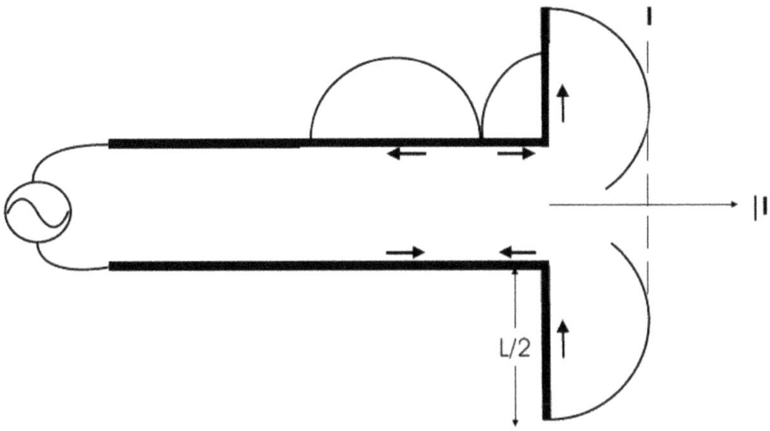

Figure 4.6. The radiation fields are cancelled in the parallel section of the wire as the current flow I represented by the arrow is in the opposite direction. In the flared section of the wire (each having a length $L/2$), the radiation fields are enhanced as the currents are in the same direction, which results in radiation. This illustrates the way that an antenna can be designed by breaking the symmetry of an electrodynamic system.

that the symmetry of the electromagnetic field has been explicitly broken, resulting in electromagnetic radiation.

It is worth mentioning that the symmetry breaking of an electrodynamic system is not limited to breaking the geometric symmetry of the system. There must be explicit symmetry breaking of the electromagnetic field, which results in electromagnetic radiation. The fact is that geometric symmetries are never fully broken, but only lowered. For example, when a transmission line's existing symmetry is broken to create a dipole antenna, its initial symmetry is broken, but the dipole antenna continues to hold some form of geometric symmetry. The key aspect of the symmetry breaking of an electrodynamic system is the transformation of the system to a form which allows symmetry breaking of the electromagnetic field in space within the localised region of the electrodynamic system. A geometrically symmetric system like a spherical metal excited by a time-varying voltage source would result in electromagnetic radiation, as under such conditions the geometric symmetry of the electromagnetic field would be broken, which can be associated with the emission of electromagnetic waves. However, if the spherical structure is enclosed within another metallic enclosure and the time-varying voltage source is applied to the two systems, there would not be any radiation from the system due to the symmetries associated with the electromagnetic field. So, the symmetry breaking of an electrodynamic system can be related to the physical geometry of the system, but it is not limited to it.

It could be argued that a system of point charges has an asymmetric pattern of electric lines of force and there is no radiation. It is worth mentioning that radiation is associated with localised symmetry breaking of the electric field around the charge centres. Here, we can apply Gauss's law in the localised region of each of the charge centres which will give spatial symmetry of the associated electric field around each charge. If we consider the system of charges as a distinct entity, symmetry can be defined in terms of the spatial and temporal variance of the system's Lagrangian. In the context of static charges, the Lagrangian does not vary with time in the localised region of the configuration of charges, hence, there is temporal symmetry and absence of non-conserved Noether current resulting in no electromagnetic radiation.

4.3 Explicit symmetry breaking and free electron lasers

The concept of radiation under explicit symmetry breaking can easily be extended to infrared and optical frequencies. The best example to illustrate the phenomenon is a tuneable laser called a free electron laser, which was invented by Meady at Stanford University. The device comprises a configuration of alternate north and south poles called a wiggler magnet (figure 4.7). Electrons are fired across the wiggler magnet and they encounter magnetic flux alternately aligned in opposite directions, which causes the electrons to oscillate in a periodic manner as a result of the Lorentz force impinged by the magnetic poles. There is bremsstrahlung radiation as a consequence of the electron acceleration [5]. The wavelength of the emitted light is expressed by, $\lambda = \lambda_L/2\gamma^2$, where λ_L is the wavelength of the undulator magnet and γ is the relativistic factor given by $1/\sqrt{(1 - v^2/c^2)}$, where v is the electron velocity and c is the speed of light.

In the case of an electron moving at a uniform velocity, the electric and magnetic field in a reference frame attached to the electron remains constant and the associated Lagrangian has translational as well as temporal invariance. As the electron is subjected to a periodically varying Lorentz force generated by its motion under the Wiggler magnet, the translational symmetry of the electric field is explicitly broken in the inertial frame attached to it, which results in the generation of a time-varying magnetic field, which eventually results in electromagnetic radiation. Apart from the symmetry breaking of the electric field in a free electron laser, the magnetic field of the electron beam also undergoes symmetry breaking. When the electron beam has a uniform velocity, the magnetic field intensity in an inertial reference frame moving with the electrons $H = I/2\pi r$, where I is current and r is radial distance from the charge in space, has translational as well as temporal symmetry. As the electron beam undergoes deflection under the impact of the alternating magnetic field, the magnetic field of the electron undergoes translational as well as temporal symmetry breaking, which is directly associated with the

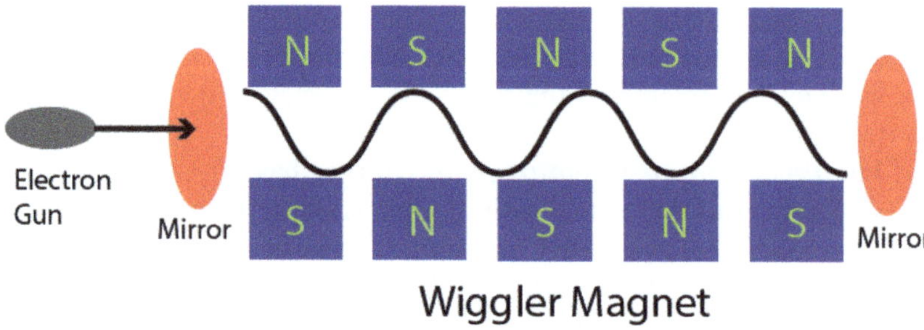

Figure 4.7. Schematic diagram of a free electron laser.

symmetry breaking of the electric field, resulting in electromagnetic radiation. The electron beam is slowed down as it loses its energy to the non-conserved Noether current that emerges from the system as electromagnetic radiation. For continuous emission of radiation, the electron needs to be energised periodically. The advantage of free electron laser is that it can be used to generate electromagnetic radiation from the infrared to the x-ray range.

4.4 Electromagnetic radiation under explicit symmetry breaking of filter circuits

A widely used electrodynamic system with inherent symmetry is an electronic filter; this behaves like a conduit for the selective transfer of energy from the input to the output side in an electrical circuit. Again, it is worth reminding the reader that symmetry is not limited to the structural geometry of the system, but is associated with the electrodynamic properties of the system with regard to the way energy transfers within the system. In the case where a pair of transmission lines is being used as a filter, there is electrical as well as geometric symmetry. However, an inductor–capacitor system can also be used as a filter where the geometric symmetry simply does not have much empirical relevance. In such cases, the symmetry is defined in terms of the invariance of the input and output terminals and the energy flow associated with it. For example, energy fed to the input terminal appears at the output terminal and vice versa. The flux of energy has translational as well as temporal invariance with regard to the system. The most important example in this context is that of a resonator which is used as a filter by connecting the device in a symmetric manner.

We shall see that the transformation of the symmetries of resonators that are symmetric filters such that the existing symmetry is broken can lead to the development of a radiating system. A resonator is a medium of energy storage comprising a set of resonant frequencies which are determined by its structural parameters. Under a symmetric form of connections or symmetric excitations, resonators work like filters, allowing the selective transfer of energy from the input to the output terminal. When the symmetry of such a system is explicitly broken by suitable structural changes, the asymmetric resonator can behave like radiating antennas. The simplest example of such a symmetry breaking is to feed excitation from one end and leave the other end floating in free space. Energy fed into one end of the system will leave the system from the other end.

Under the conditions of symmetry, the system's Lagrangian shows temporal as well as spatial invariance; when the symmetry is explicitly broken, the associated Lagrangian has terms which explicitly make it non-invariant [1]. The theoretical perspective can be used to explore explicit symmetry breaking at the empirical level in different kinds of systems. A simple example of a resonator under symmetric excitation acting like a filter is a capacitor–inductor circuit with one end connected to a source of electrical excitation and the other end connected to the ground terminal. In such a system, electrical charges oscillate between the capacitive and inductive elements as a result of thermal excitations. The system has a precise set of resonant frequencies, which allows it to selectively transfer energy from the excitation source to the ground. The Lagrangian of the system has temporal invariance as the energy from the excitation source is transferred to the capacitive element and is eventually absorbed as the kinetic energy of electrons in the inductor and is circulated back into the system via the ground terminal in a periodic manner. Assuming that the capacitor and inductor are idealized lumped elements confined to certain points in space, the determination of the spatial dependence of the invariance of the Lagrangian is not physically possible in the current context.

The concept can be generalized to describe the fact that all resonators under symmetric conditions of excitation act like filters which selectively transfer energy from the input to the output terminal. Such symmetric resonators or filters exhibit translational and temporal symmetries as the flux of energy within the device is homogenous with the change in spatial and temporal coordinates. Similarly, the Hamiltonian and the Lagrangian densities during translational or temporal variations within the system remain constant. In addition, the system's response remains invariant to an interchange of its input and output ports or leads connected to them.

Explicit symmetry breaking of the existing symmetry of filters would result in variation of the existing energy flux, which would cause either storage or dissipation of energy. For example, when the excitation is fed to the input terminal and the output terminal is closed, the energy is fed back towards the input terminal. Similarly, when the output terminal is opened or left floating in free space, energy is radiated out from the system. The radiation will be relatively higher around the system's resonant frequencies.

A parallel two-wire transmission line fed with excitation at one end is a symmetric resonator which can act like a filter, selectively transferring signals from the input to the output. The spatial and temporal invariance of its Lagrangian density have been discussed before. Its explicit symmetry breaking would involve creating changes such that the associated Lagrangian loses its temporal as well as spatial invariance. In the simplest terms, this can be done by flaring the end terminal of the transmission line, which breaks the symmetry of the electric lines of field, which is also reflected in the symmetry breaking of the Lagrangian along the temporal as well as the spatial dimension. Thus, the resonator-like properties of a transmission line under symmetric excitation can be used to make it a filter, however, it can be transformed into an antenna by breaking the symmetry of the resonator. The concept can be generalised and we can say that antennas can be created by explicitly breaking the symmetry of resonators by applying asymmetric excitation.

An electrical resonator made of a set of inductors is shown in figure 4.8. Its symmetry is explicitly broken as the output terminals are left floating and excitation is fed to the input side. We can call such a means of excitation, asymmetric excitation. The asymmetric filter circuit no longer remains a selective conduit of energy transfer between the input and output terminals. The energy fed to it from the input side is stored and subsequently radiated out to free space. Thus, the asymmetric resonator in the current context works like an antenna. The concept is rather broad and general and can be applied to all electrical resonators which can be transformed into antennas under asymmetric excitation.

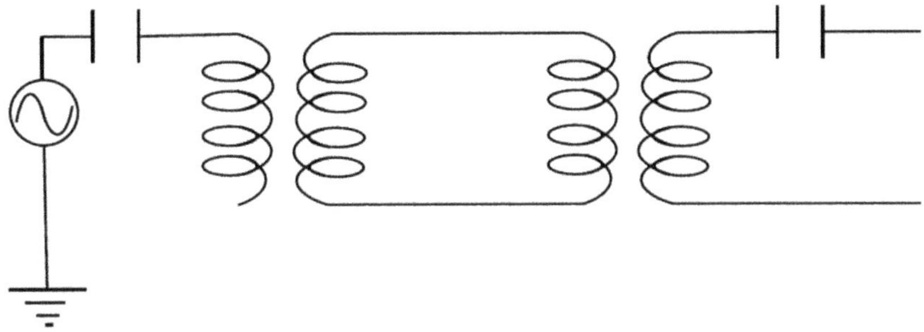

Figure 4.8. An LC circuit under asymmetric excitation where one of the terminals is excited and the other is left floating in free space.

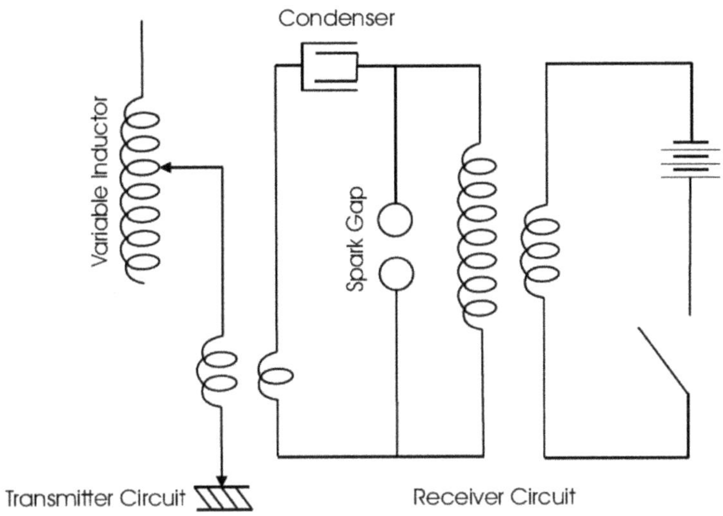

Figure 4.9. Schematic diagram of the antenna invented by Marconi, comprising an inductor floating in free space with energy fed under electromagnetic induction [6].

Some of the early work in the field of antennas was based on the concept of the symmetry breaking of resonators. In fact, the initial observations on the relationship between symmetry breaking and radiation were made by Guglielmo Marconi in a rather intuitive manner; he realised that putting an aerial into a vertical position increased its ability to receive and transmit signals. From a physical perspective, Marconi broke the symmetry of the electrodynamic system by aligning the antenna in a vertical direction. The concept is very general and can be applied to all kinds of electrodynamic systems where a transformation of the existing symmetry that breaks it in some form can result in electromagnetic radiation.

Figure 4.9 shows the schematic diagram of an antenna invented by Marconi in the last years of the 19th century. The system comprises a transmitting antenna with an inductor with one terminal floating in free space while the other terminal is connected to an inductor wired to the ground, and this receives time-varying electrical excitation from another circuit through electromagnetic induction. Unlike a typical electrical filter, where the electrical energy traverses the system and flows back to the ground or is transferred to the output terminal, here, the input energy is transmitted into free space. Thus, in all antennas, the symmetry breaking of the system results in temporal and spatial invariance of the associated Lagrangian and electromagnetic radiation results from the loss of the Noether current from the system.

The concept of symmetry breaking in simple electrodynamic systems like transmission lines and inductor–capacitor systems offers a new perspective in antenna design. Although the ideas have been used in communication systems for more than a century, a theoretical framework for the correlation between efficient radiators and electrodynamic asymmetry was missing until recently. In addition to the empirical applications of the idea in antenna design, it also has theoretical importance, as it supports the viewpoint that electron acceleration is not sufficient to generate electromagnetic radiation—there must be explicit symmetry breaking of the electromagnetic field.

In the next chapter, the concept is applied to explain the workings of antennas made of dielectric materials. The aim is to resolve a theoretical question on the operation of dielectric antennas that has been debated for more than 50 years.

References

[1] Brading K and Castellani E 2003 *Symmetries in Physics: Philosophical Reflections* (Cambridge: Cambridge University Press)

[2] Sinha D and Amaratunga G A J 2015 Electromagnetic radiation under explicit symmetry breaking *Phys. Rev. Lett.* **114** 147701

[3] Balanis C A 1982 *Antenna Theory, Analysis and Design* (New York: Harper and Row)

[4] Noether E 1918 *Nachr. D. König. Gesellsch. D. Wiss. Zu Göttingen, Math-phys. Klasse* 235–57

[5] Deacon D A, Elias L R, Madey J M, Ramian G J, Schwettman H A and Smith T T 1977 *Phys. Rev. Lett.* **38** 892

[6] Marconi G 1900 *GB Patent Specification* 7777

IOP Publishing

Explicit Symmetry Breaking in Electrodynamic Systems and Electromagnetic Radiation (Second Edition)

Dhiraj Sinha and Gehan A J Amaratunga

Chapter 5

Symmetry breaking in transformation of force fields

5.1 Introduction

In the previous chapter, we looked at how symmetries in physical systems are associated with conserved currents and the system's Lagrangian. We also saw how the non-conserved Noether current emerging from a localized spatial and temporal domain of a radiating system is transformed into electromagnetic radiation as a consequence of the broken symmetry of the electromagnetic field within the system. In the current chapter, we will see how some of these are interwoven into the very nature of Maxwell's equations and Gauge symmetry, which is related to the global conservation of charges. The loss of charge from a system implies loss of Noether current at the local level and localized symmetry breaking. Hence, it can be seen that global symmetries and global conservation of charges as well as locally broken symmetries are embedded into the framework of Maxwell's equations.

5.1.1 Localized symmetry breaking

The Noether theorem establishes a link between symmetries and conserved currents [1], however, it does not say anything in specific terms about the broken symmetries and non-conserved currents which can define the properties of force fields. Such issues can become important in the context of transformation of force fields and related constraints. To highlight this aspect, we consider, a body of mass m, symmetrically placed in a gravitational field (figure 5.1), with a Lagrangian $\mathcal{L}$ which has a spatial symmetry along its spatial dimension r, defined as $d\mathcal{L}/dr = 0$ [2]. When it is displaced away from its equilibrium position, the symmetry of the Lagrangian is

doi:10.1088/978-0-7503-5132-4ch5

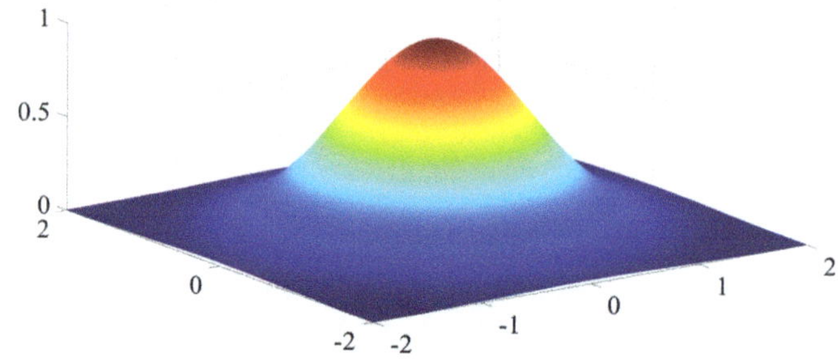

Figure 5.1. Symmetry breaking. A mass placed at the crest of the energy surface does not see any force field due to symmetry. A displacement along any direction breaks its symmetry within a localized region, which imparts momentum. Momentum is conserved globally but is lost within a local region where its symmetry is broken. Reproduced with permission from [5], copyright The Royal Society.

explicitly broken within a localized region of space and $d\mathcal{L}/dr \neq 0$. This is also defined through the Euler–Lagrange equation,

$$\frac{d}{dt}\left(\frac{d\mathcal{L}}{d\dot{r}}\right) = \frac{d\mathcal{L}}{dr} \tag{5.1}$$

where $\dot{r}$ indicates velocity. It essentially implies that force is equal to rate of change of momentum, i.e. $d\mathcal{L}/dr = d\boldsymbol{p}/dt$, where $\boldsymbol{p}$ is momentum and t is time. The momentum $\boldsymbol{p}$ varies along the temporal dimension, i.e. $d\boldsymbol{p}/dt \neq 0$, it could be argued that its symmetry along the temporal dimension is locally broken. It implies that local symmetry breaking of potential along spatial dimension is associated with localized symmetry breaking of momentum along the temporal dimension, which maintains global symmetry of the system.

The relationship between temporal and spatial symmetry breaking of specific fields can also be extended to electromagnetism. If the Lagrangian of a charge of magnitude e is $\mathcal{L}_\mathcal{E}$, then $d\mathcal{L}_\mathcal{E}/dr \neq 0$ implies that the particle is under the influence of an electric field $\boldsymbol{E}$ and $d\mathcal{L}_\mathcal{E}/dr = e\boldsymbol{E}$. As the charge accelerates, the spatial symmetry of the electric lines of field is broken and a rotational electric field is generated expressed as, $\nabla \times \boldsymbol{E}$ which is associated with a time varying magnetic flux density $\boldsymbol{B}$ such that $\nabla \times \boldsymbol{E} = -\partial \boldsymbol{B}/\partial t$. For a plane polarized electromagnetic wave with its velocity vector along the z axis, the electric and magnetic field lines are oriented along x and y directions, respectively, and the corresponding Maxwell's equation is [3],

$$\frac{\partial E_x}{\partial z} = \frac{\partial B_y}{\partial t} \tag{5.2}$$

It essentially signifies that the translational symmetry breaking of the electric field intensity along a spatial dimension is linked to the temporal symmetry breaking of the magnetic flux density, which generates radiation [4]. It shows how electric and magnetic field vary along spatial and temporal dimensions in the context of electromagnetic radiation. In the given equation, $\partial E_x/\partial B_y$ also indicates how the space is curved. In the context of a flat space, the electric and magnetic fields are aligned perpendicular to each other and the electromagnetic wave traverses a straight line [5]. This orthogonality is transformed in the context of curved spaces. Thus, the concept of localized symmetry breaking is embedded within the existing formulation of Maxwell's equation and it demonstrates how the force fields along spatial and temporal dimensions are interdependent. For radiation, the spatial as well as temporal symmetry of the Lagrangian should be broken. For example, in a static configuration of two charges, as shown in figure 5.2, there is no radiation as the Lagrangian does not change along temporal dimension and localized non-conserved Noether current is not generated, although the field lines have an asymmetric profile.

By using Stoke's theorem in the expression $\nabla \times \boldsymbol{E} = -\partial \boldsymbol{B}/\partial t$, we can write,

$$\int \boldsymbol{E} \cdot d\boldsymbol{r} = -\frac{\partial \int \boldsymbol{B}.\, d\boldsymbol{S}}{\partial t} = -\frac{d\phi}{dt} \tag{5.3}$$

where $\boldsymbol{S}$ denotes surface area vector and ϕ represents the magnetic flux density. Here, $d\phi/dt$ expresses symmetry breaking of magnetic flux lines along the temporal dimension. The electric field on the left hand slide of the expression is symmetry breaking of electric potential along spatial dimensions, which generates an electric field vector in space whose integral is electric potential. Thus, the expression, which defines Faraday's law of induction also expresses a correlation between temporal

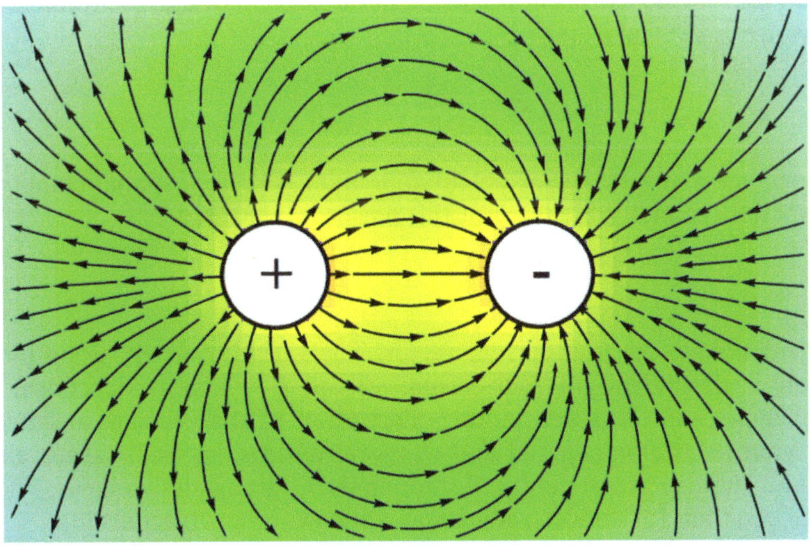

Figure 5.2. Electric field lines of a dipole. In an electric dipole, the electric lines of field have an asymmetric profile but there is no radiation. Radiation is generated when symmetry is changed along spatial as well as temporal dimension. Reproduced with permission from [5], copyright The Royal Society.

and spatial symmetry breaking of field in space and transformation of force fields. The expression defined in equation (5.3) is usually discussed in the context of voltage induction in metallic coils under time varying flux [6] but also holds for insulating materials like piezoelectric materials [7].

The relationship between the geometric orientation of magnetic field intensity and electric flux density, for an electromagnetic wave propagating in free space, is represented as [3],

$$\nabla \times \boldsymbol{H} = \frac{\partial \boldsymbol{D}}{\partial t} \tag{5.4}$$

where $\boldsymbol{H}$ represents magnetic field intensity and $\boldsymbol{D}$ is electric flux density. For a specific components of these vector fields in space in the context of a plane propagating wave, we can write,

$$\frac{\partial H_y}{\partial z} = \frac{\partial D_x}{\partial t} \tag{5.5}$$

where H_y is magnetic field intensity along the y direction and D_x is electric flux density along the x direction. The equation essentially expresses the fact that the symmetry breaking of electric flux density along the temporal dimension is associated with symmetry breaking of magnetic field intensity along spatial dimension. Thus, the temporal and spatial dependence of force fields during their transformation and locally broken symmetries are fundamentally expressed in the basic laws of electrodynamics. The Noether currents are locally non-conserved but are globally conserved [5].

We can also extend the prior discussion in the context of the electromagnetic field tensor, which is defined as [8],

$$\boldsymbol{F}_{\mu\nu} = \frac{\partial \boldsymbol{A}_\nu}{\partial r_\mu} - \frac{\partial \boldsymbol{A}_\mu}{\partial r_\nu} \tag{5.6}$$

Here $\boldsymbol{A}$ denotes magnetic vector potential, μ and ν represent spatial components of $\boldsymbol{A}$. The spatial symmetry breaking of this tensor is expressed as,

$$\frac{\partial \boldsymbol{F}_{\mu\nu}}{\partial r} = \boldsymbol{J}_\mu \tag{5.7}$$

where $\boldsymbol{J}_\mu$ represents the four-current vector. The variation of $\boldsymbol{F}_{\mu\nu}$ along the spatial dimension is dependent on four-current vector $\boldsymbol{J}_\mu$ which is dependent on the temporal dimension. It is yet another illustration of the correlation between the broken symmetries along spatial and temporal dimensions and transformation of force fields. This aspect also finds reflection in the Lagrangian dynamics. The Euler–Lagrange equation for an electrodynamic system can be written as [8],

$$\partial_\beta \left[\frac{\partial \mathcal{L}_D}{\partial(\partial_\beta A_\alpha)} \right] - \frac{\partial \mathcal{L}_D}{\partial A_\alpha} = 0 \tag{5.8}$$

Here, $\mathcal{L}_\mathcal{D}$ represents the Lagrangian density with subscripts, α and β, denoting its components in space. The expression, $\partial\mathcal{L}/\partial A_\alpha = -J^\alpha$, represents the broken symmetry along the spatial domain generating a Noether current [5].

As radiation emerges from a specific region in space, the associated Lagrangian loses its temporal symmetry within the localized region. With time, the wave moves away from the given region and the value of Lagrangian density becomes zero within that region. Thus, the global symmetry of the Lagrangian is maintained, despite the fact that it is broken, within a localized region of space.

Symmetry of a physical system is a relative quantity and is defined in terms of invariance of a set of physical quantities which are said to form a transformation group, referred to as the symmetry group [9]. Symmetry breaking essentially defines the transformation relations among a given symmetry group and its sub-groups, where the initial symmetry is transformed into one of its sub-groups.

Translational or rotational symmetry breaking of geometric lines of a given field is associated with a change in order of symmetry of the given field. To put this idea in perspective, we consider a ferroelectric material with electric dipole moments oriented along random geometric directions. The net dipole moment of the system remains invariant to geometric rotation, i.e. any kind of measurement on the electric dipole moment would lead to a net value of zero, we can say that the electric dipole moment has rotational symmetry. When the material is subjected to an electric field, a net polarization is induced which can be written as $\boldsymbol{P_e} = \chi_e \epsilon \boldsymbol{E}$, where χ_e represents electric susceptibility and ϵ denotes the electric permittivity [10]. We can say that the rotational symmetry of polarization field has been explicitly broken. However, the process has generated a new symmetry, which is the translational symmetry of the polarization field. However, radiation would not emerge from such a system as the Lagrangian has symmetry with regard to temporal variation. Thus, a specific instance of symmetry breaking can induce new symmetries and symmetries are not totally lost in such transformations.

The generation of curved space in the general theory of relativity can also be considered from the perspective of symmetry breaking. Mass explicitly breaks the geometric symmetry of space–time fabric by changing its existing curvature (figure 5.3(a)) [5]. The algebraic relationship is expressed as [11],

$$\boldsymbol{G}_{\mu\nu} = \boldsymbol{R}_{\mu\nu} - \frac{1}{2}\boldsymbol{R}\boldsymbol{g}_{\mu\nu} = \frac{8\pi}{c^4}G\boldsymbol{T}_{\mu\nu} \tag{5.9}$$

In this equation, c is the speed of light, $\boldsymbol{G}_{\mu\nu}$ is Einstein tensor and represents the curvature of space–time surface, $\boldsymbol{T}_{\mu\nu}$ is the energy momentum tensor which quantifies the total kinetic energy of the system. In the absence of mass, $\boldsymbol{R}_{\mu\nu} = 0$ and the space–time surface has zero curvature. The tensor $\boldsymbol{G}_{\mu\nu}$ represents the potential energy gradient which can be linked with spatial symmetry breaking of a field and $\boldsymbol{T}_{\mu\nu}$ can be associated with temporal symmetry breaking of the Noether current. It is graphically illustrated in figure 5.3(b), where the crests and valley in the space–time surface around a massive object is illustrated.

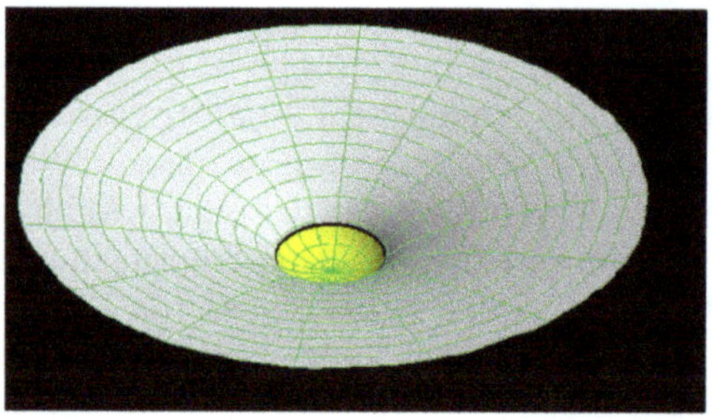

(a)

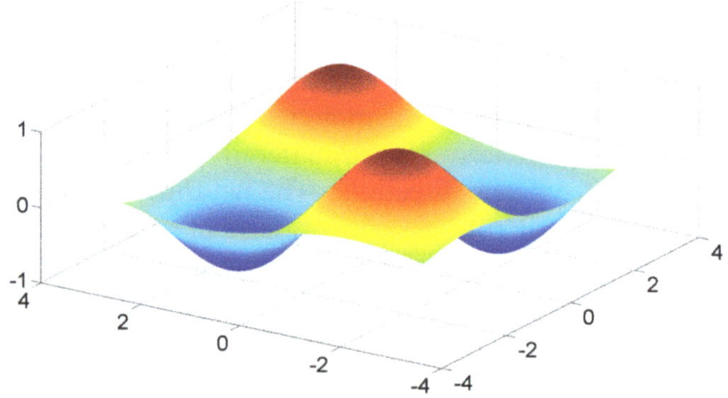

(b)

Figure 5.3. Explicit symmetry breaking of the curvature of space–time surface. (a) The presence of mass breaks the symmetry of space–time curvature. (b) Space–time surface always bends around mass. Reproduced with permission from [5], copyright The Royal Society.

5.2 Broken symmetry in electrodynamic systems, Gauge symmetry and radiation

Global symmetry in electrodynamic systems indicates charge conservation which is defined using gauge invariance [12]. The electric and magnetic fields vectors are linked to a physical quantity called magnetic vector potential, which is expressed by the relationship [8],

$$\boldsymbol{E} = -j\frac{\nabla\nabla \cdot \boldsymbol{A}}{\omega\mu\epsilon} - \frac{\partial A}{\partial t} \tag{5.10}$$

where μ denotes permeability, ω is frequency and j is iota. The magnetic flux density $\boldsymbol{B}$, can be expressed in terms of the magnetic vector potential [8],

$$\boldsymbol{B} = \nabla \times \boldsymbol{A} \tag{5.11}$$

The equation indicates that magnetic flux density is equal to rotation of the magnetic vector potential. In other words, the translational symmetry breaking of magnetic vector potential generates magnetic flux density.

According to the principle of Gauge invariance, the field $\boldsymbol{B}$ remains invariant to changes in $\boldsymbol{A}$. For example $\boldsymbol{A} = \boldsymbol{A} + \nabla\psi$, where ψ is a scalar field, leads to the invariance of magnetic field and $\boldsymbol{B} = \nabla \times (\boldsymbol{A} + \nabla\psi) = \nabla \times \boldsymbol{A}$ for a given physical system [8]. We can argue that the rotational symmetry breaking of the magnetic vector potential remains invariant to finite changes in it. In other words, the magnetic flux density is an outcome of translational symmetry breaking of the magnetic vector potential, hence, it is not influenced by changes in its existing symmetry along a specific spatial dimension.

Gauge invariance also sets some conditions on the electric field intensity. For example, $\boldsymbol{A} = \boldsymbol{A} + \nabla\psi$, where ψ is a scalar field, leads to the invariance of $\boldsymbol{E} = -\nabla V - \partial \boldsymbol{A}/\partial t$. Another key aspect of Gauge invariance is that a change in $\boldsymbol{A}$ by an amount $\nabla\psi$ leads to a corresponding decrease in the electric potential $V \rightarrow V - \partial\psi/\partial t$ [19, 20]. The electric field of a uniform plane wave is $\boldsymbol{E} = -\partial \boldsymbol{A}/\partial t$, which essentially implies that symmetry breaking of the magnetic vector potential along temporal dimension generates an electric field, which is invariant to a change in $V \rightarrow V - \partial\psi/\partial t$. Hence, the invariance of $\boldsymbol{E}$ and $\boldsymbol{B}$ is ensured by the localized symmetry breaking of the magnetic vector potential $\boldsymbol{A}$, along spatial and temporal dimensions. This further defines the invariance of the Lagrangian describing the electrodynamic system.

Here, we have established a link between the broken symmetries of electric and magnetic fields along temporal and spatial dimensions in terms of magnetic vector potential and its relationship to gauge invariance. We have explained how the nature of symmetry breaking of magnetic vector potential within a localized region leads to gauge invariance and symmetries at a global scale [5].

5.3 Symmetry breaking and reversibility

Probing the nature of irreversibility in the context of symmetry breaking opens a novel framework of exploration. For example, whether spontaneous and explicit symmetry breaking fall within a reversible or an irreversible process is an intriguing question. In order to understand the nature of irreversibility, the key role of broken symmetries and associated Noether currents needs to be taken into account. The fundamental laws of physics have time reversal symmetry, but irreversibility is embedded within the framework of real physical systems.

The question of irreversibility has been investigated from a statistical notion since the days of Boltzmann [13]. It is defined from the perspective of evolution of the number of microstates which increases the accessible states and an increase in entropy. However, there is no mathematical framework to irreversibility beyond a statistical framework. For example, Brian Greene writes, 'No one has ever discovered any fundamental law which might be called the Law of the Spilled Milk or the Law of the Splattered Egg' (*The Fabric of the Cosmos: Space, Time and the Structure of Reality*) [14]. Lebowitz has left a note on Scholarpedia, 'It is only secondary laws, which describe the behaviour of macroscopic objects containing many, many atoms, such as the second law of thermodynamics, which explicitly contain this time asymmetry. The obvious question then is; how does one go from a time symmetric description of the dynamics of atoms to a time asymmetric description of the evolution of macroscopic systems made up of atoms' [15].

The phase space of a system increases during entropy increase in an isolated system [16]. A hot cup of coffee cooling down spontaneously is a related example (figure 5.4). According to Liouville's theorem of classical mechanics, the volume of phase space is conserved [17]. An increase in entropy with the evolution of microstates which leads to irreversibility does not agree with the time reversal symmetry which is fundamental to Newtonian mechanics. This is called the Loschmidt paradox [18]. A statistical explanation is offered based on the coarse graining of phase space, which has some key theoretical issues [19]. Currently, a mathematical model on irreversibility using a deterministic framework does not exist. In a recent work, this aspect has been addressed using the geometric trajectories of Noether current.

Transformation of force fields is a reversible process. For example, potential energy can be transformed into kinetic energy and vice versa. In such systems, a change along a spatial coordinate system leads to a change of a corresponding physical quantity along the temporal dimension, which is expressed in the equation, $dU_m/dr = dp/dt$. A similar kind of reversibility is observed in the context of electromagnetism where electric and magnetic fields can be transformed into each other such that a change along a spatial dimension of a physical quantity is associated with a change in some quantity along the temporal dimension. The Maxwell's equations, $\nabla \times \boldsymbol{H} = \partial \boldsymbol{D}/\partial t$ and $\nabla \times \boldsymbol{E} = -\partial \boldsymbol{B}/\partial t$, which establish a relationship between temporal and spatial variations of electric and magnetic fields,

Figure 5.4. Irreversibility in a thermodynamic system. Spontaneous cooling of a hot cup of coffee is an instance of irreversibility.

essentially describe this key aspect of physical reality. In such systems, we can define a specific surface through which a globally conserved Noether current traverses, which can be linked with global symmetry and reversibility. In a thermodynamic system whose microstates are evolving with time spontaneously, we cannot define a conserved Noether current [5]. The corresponding physical quantities lack a linear dependence on spatial and temporal coordinates as in the context of electromagnetic fields.

In the context of heat transfer [16],

$$\nabla^2 T = \frac{1}{K_T}\frac{\partial T}{\partial t} \tag{5.12}$$

where K_T represents thermal conductivity and T is temperature. The first term in the equation $\nabla^2 T$, defines the value of the term on the right hand side (RHS) of equation, $\partial T/\partial t$. However, the RHS term, which is dependent on time does not lead to the term on the left which has a spatial dependence. Thus the spatial dependence of temperature and its temporal dependence do not have the kind of relationship and interdependence which is found in processes having global symmetry having a well defined conserved current.

Now, we can probe the prior question on the nature of reversibility in symmetry breaking. In a reversible process, changing the direction of flow of energy brings the system back to its initial state. For example, heat generation under work transfer and an increase in entropy is an irreversible process as reversing the direction of work does not reduce the entropy of the system. Cooling down of a hot cup of coffee is an illustrative example of entropy increase of the Universe.

Transformation of water into ice and induction of ferromagnetism in a ferromagnetic material below the Curie temperature are instances of spontaneous symmetry breaking [20]. In such cases, heat is removed from the system leading to a decrease in entropy of the system. A set of gas molecules could be heated and cooled using a reversible process and the process can be called reversible [16]. For example, reversible isothermal expansion and compression is one such example. Considering this, removing heat from a system, subsequent phase transition and spontaneous symmetry breaking and then heating it up to bring it back could be considered an instance of a reversible process, where the entropy change of the system and the surrounding is zero. However, if we consider the non-equilibrium aspects of the process, it would appear that the process is irreversible. When heat is removed from the system, it would entail a thermodynamic cycle, which in itself is not reversible. Heat addition will bring the system to its initial equilibrium. In the context of a ferromagnetic material, if heat is added to the system, it will revert back to its initial state of zero magnetism, but it changes the entropy of the system as well as the surroundings. If heat is removed again, the system's final alignment of magnetic moments could be different from the initial configuration. Thus, it could be argued that spontaneous symmetry breaking is not a reversible process if one considers the dissipative aspects of the process.

The nature of reversibility in explicit symmetry breaking also needs special attention. It is of special interest from the perspective of a historical debate on irreversibility associated with generation of a photon. Einstein was of the view that emission of a photon is an irreversible process [21]. His main argument was that spontaneous absorption of a photon is never observed. Einstein's idea could be easily extended to electromagnetic emission and it could be argued that it is an irreversible process as the waves tend to spread out and they are never absorbed by the emitting device again. Planck believed that emission of a photon is a reversible process [22].

A theoretical analysis while considering the nature of force fields and the nature of transformation of force fields and their temporal and spatial dependence indicates that emission of an electromagnetic wave is a reversible process. The space and time are correlated through fields which are dependent on each other for their transformation and there is a specific Noether current, which is conserved globally. While reversing the physical variables, we can reverse the whole process and bring the entire system back to its initial configuration without generation of heat, which further highlights the reversible nature of symmetry breaking.

In the current chapter, we have seen the broader applicability of explicit symmetry breaking in the context of transformation of force fields. The symmetry of a force field is broken along spatial dimension and it is associated with a corresponding symmetry breaking of a related field along the temporal dimension. The Noether current is globally conserved and locally non-conserved. In electromagnetism, the symmetries of electric and magnetic fields are broken along spatial dimensions which is linked to temporal changes of corresponding fields. The process is integrated with generation of radiation, where Noether current is lost within a localized region of space.

Some of the ideas on broken symmetries and transformation of force fields can also be applied in the context of understanding the nature of irreversible processes. In a non-equilibrium thermodynamic process, where heat is transferred from a higher temperature to a lower temperature, leading to an increase in entropy, the interdependence of associated physical quantities on space and time do not have a linearity, which is found in electromagnetism. Thus, we can define irreversibility using a closed form mathematical expression by considering their symmetry variations along spatial and temporal dimensions. These ideas can also be used to prove that emission of electromagnetic radiation is, in fact, a reversible process.

References

[1] Noether E 1918 Nachr. D. König. Gesellsch. D. Wiss. Zu Göttingen *Math-phys. Klasse* **325**

[2] Goldstein H 1957 *Classical Mechanics* (Boston, MA: Pearson)

[3] Hayt W H 1981 *Engineering Electromagnetics* (New York: McGraw-Hill)

[4] Sinha D and Amaratunga G A J 2015 Electromagnetic radiation under explicit symmetry breaking *Phys. Rev. Lett.* **114** 14770

[5] Sinha D and Amaratunga G 2018 The Noether current in Maxwell's equations and radiation under symmetry breaking *Phil. Trans. R. Soc. A* **376** 20170452

[6] Galili I, Kaplan D and Lehavi Y 2006 Teaching Faraday's law of electromagnetic induction in an introductory physics course *Am. J. Phys.* **74** 337–43

[7] Sinha D, Pisana S and Flewitt A J 2011 Radio frequency magnetic field detection using piezoelectric coupled microcantilevers *Smart Mater. Struct.* **20** 025016
[8] Jackson J D 1988 *Classical Electrodynamics* (Amsterdam: Wiley)
[9] Hamermesh M 1989 *Group Theory and its Application to Physical Problems* (North Chelmsford, MA: Courier Corporation)
[10] Resta R 1992 Theory of the electric polarization in crystals *Ferroelectrics* **136** 51–5
[11] Carroll S M 2004 *Spacetime and Geometry: An Introduction to General Relativity* (Boston, MA: Addison-Wesley)
[12] Quigg C 2013 *Gauge Theories of the Strong Weak, and Electromagnetic Interactions* (Princeton, NJ: Princeton University Press)
[13] Brown H R, Myrvold W and Uffink J 2009 Boltzmann's H-theorem, its discontents, and the birth of statistical mechanics *Studies in History and Philosophy of Science Part B: Studies In History and Philosophy of Modern Physics* vol 40 (New York: Elsevier) pp 174–91
[14] Greene B 2007 *The Fabric of the Cosmos: Space, Time, and the Texture of Reality* (New York: Alfred Knopf)
[15] Lebowitz J L 2008 Time's arrow and Boltzmann's entropy *Scholarpedia* **3** 3448
[16] Reif F 1965 *Fundamentals of Statistical and Thermal Physics* (New York: McGraw-Hill)
[17] Gibbs J W 1984 On the fundamental formula of statistical mechanics, with applications to astronomy and thermodynamics *Proc. Am. Assoc. Adv. Sci.* **33**
[18] Loschmidt J 1876 *Sitzungsber. Kais. Akad. Wiss. Wien, Math. Naturwiss. Classe* **73**
[19] Vladislav C and Sheehan D P 2005 *Challenges to the Second Law of Thermodynamics* (Dordrecht: Springer)
[20] Fröhlich J, Simon B and Spencer T 1976 Phase transitions and continuous symmetry breaking *Phys. Rev. Lett.* **36** 804
[21] Einstein A 1905 Generation and conversion of light with regard to a heuristic point of view *Ann. Phys.* **17** 132–48
[22] Planck M 1991 *The Theory of Heat Radiation* (New York: Dover)

IOP Publishing

Dhiraj Sinha and Gehan A J Amaratunga

Chapter 6

Piezoelectric antennas

The phenomenon of electromagnetic radiation under explicit symmetry breaking was further explored using thin films of piezoelectric material, which were tested for their ability to generate electromagnetic waves under asymmetric electrical excitation. A sample of thin films of piezoelectric material comprising two sets of interdigital electrodes was mounted on a ground plane. The first set of electrodes were used to input electrical signals which were transformed into a surface acoustic wave (SAW) and propagated along the surface of the piezoelectric film. When the SAW reached the other set of interdigital electrodes, it was transformed back into an electrical signal. This arrangement provides a selective filtering of the input signal as the device can only support a certain bandwidth of signals while providing a very high level of attenuation to other signals [1]. In terms of electrical characteristics, the SAW filter closely resembles a typical resistor–inductor–capacitor filter due to the finite values of resistance, inductance and capacitance associated with the thin film and the electrodes [2]. Such SAW devices are widely used in televisions and mobile phone handsets as filters, but they are connected in a symmetric manner, i.e. the flux of energy across all the elements of the filter remains constant along temporal as well as spatial dimensions.

SAW filters mounted on a ground plane were electrically excited in an asymmetric mode where the upper set of electrodes were connected together in order to create a long monopole antenna-like structure. Figure 6.1 illustrates the electrode layout diagram of such a SAW device. The input as well as the output ground electrodes of the SAW filter were left floating in free space. Unlike the symmetric excitation of the SAW filter, where the input signal fed to the input set of electrodes travels to the output electrodes and is collected to be transferred to the connecting elements, here, the SAW device behaves like an asymmetric resonator, storing the input energy and feeding it back to the ground plane. The transfer of electrons between the SAW device with a capacitive reactance and the ground plane with an inductive reactance provides a framework for the explicit symmetry breaking of the electric field, which results in the generation of electromagnetic waves [3]. Under further investigation,

doi:10.1088/978-0-7503-5132-4ch6

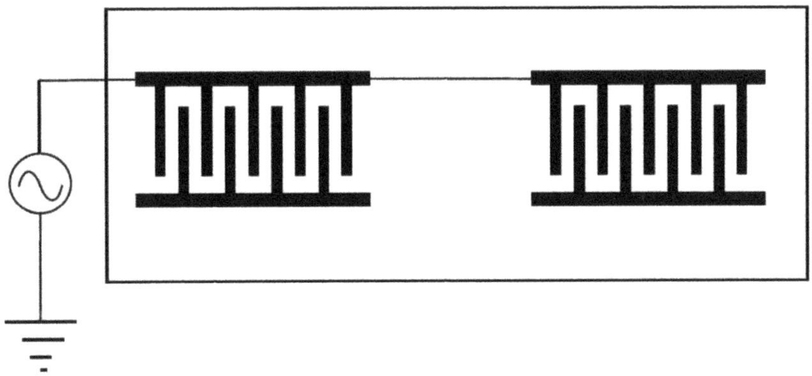

Figure 6.1. A SAW device has a thin layer of piezoelectric material along with two sets of interdigital electrodes. Under symmetric excitation, the input set of electrodes generate surface acoustic waves which are collected by the output electrodes. Here, the upper electrodes have been excited and the lower electrodes, i.e. the ground electrodes, are left floating in free space in order to create an antenna.

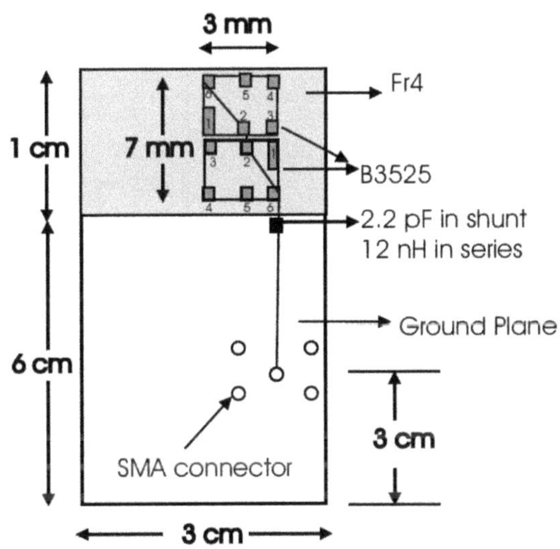

Figure 6.2. A set of two SAW filters from Epcos (total dimensions of 7 mm × 3 mm × 1 mm) was soldered onto an FR4 board with suitable impedance matching elements. Reprinted with permission from [3], copyright (2015) by the American Physical Society.

it was found that the radiation efficiency of the SAW device was dependent on the input impedance as well as the size of the ground plane. For radio frequencies where the size of the SAW device was much smaller than the wavelength of radiation, increasing the size of the ground plane increased the radiation efficiency.

The initial tests were performed using the SAW filters used for GPS applications from Epcos (B3525); they have a good return loss around 1575 MHz. A set of two SAW filters (dimensions 7 mm × 3 mm × 3 mm) was soldered onto a board (7 cm × 3 cm) made of FR4 with metallic layers that worked as the ground plane (figure 6.2). The schematic

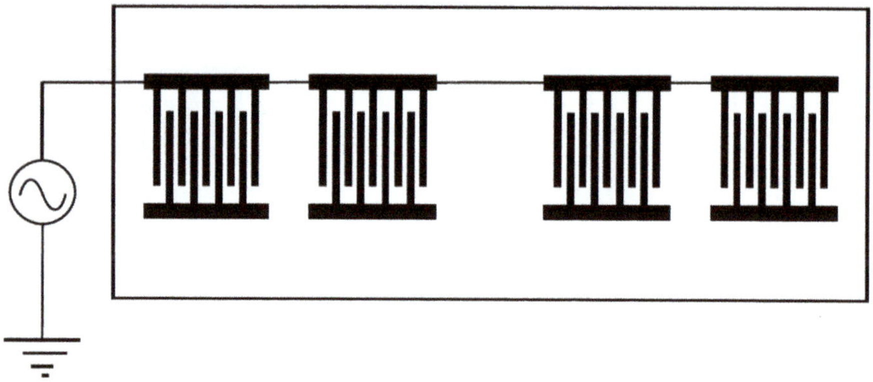

Figure 6.3. The internal electrode layout diagram for two sets of SAW filters excited in an asymmetric configuration.

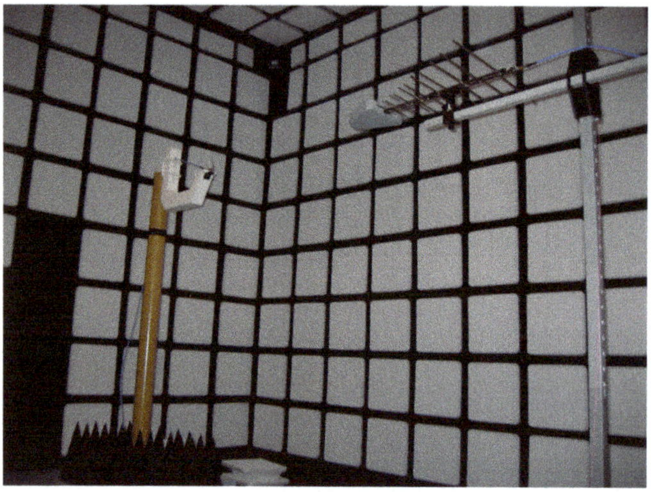

Figure 6.4. Picture of a typical anechoic chamber with piezoelectric antenna placed at a distance of 1.8 m from a test antenna. (Photo by D Sinha).

diagram representing the electrode layout is shown in (figure 6.3). The fundamental objective behind using a set of two SAW devices in series was to increase the aperture of the piezoelectric film. The radiation efficiency measurements were performed in an anechoic chamber (Satimo, Stargate 64) owned by Antenova Ltd, a Cambridge, UK-based dielectric resonator antenna company. The chamber was electromagnetically shielded using layers with high attenuation. Figure 6.4 shows the picture of one such anechoic chamber used in the tests.

Under asymmetric excitation, where the SAW filter's ground electrode is left floating, the SAW filter's impedance is capacitive in nature. As the source of voltage excitation to the antenna had an impedance of 50 Ω, the SAW device was connected to a shunt capacitor of 2.2 pF and an inductor in series to it with a value of 12 nH, which raised its impedance to 50 Ω around at 1575 MHz.

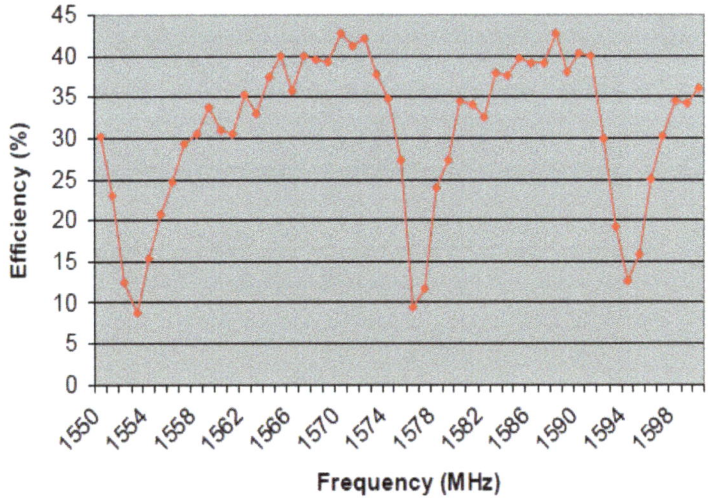

Figure 6.5. The radiation efficiency was measured to be above 40% at GPS frequencies. Reprinted with permission from [3], copyright (2015) by the American Physical Society.

Measurements of the radiation efficiency of the SAW filter as antennas are shown in figure 6.5, which shows an efficiency of 42% at 1575 MHz. The impedance matching element only works over a finite frequency range, which resulted in a drop in radiation efficiency at frequencies around 1554 MH, 1578 MHz and 1594 MHz. It is important to identify the exact role of the SAW component in radiation. Hence, the SAW filter was removed from the antenna board, leaving behind the exciting electrode. Its impedance matching values were adjusted to 50 Ω at 1590 MHz, which resulted in a monopole antenna of length 7 mm whose radiation efficiency was measured to be 22% (figure 6.6). The 20% drop in radiation efficiency validates the argument that the SAW filter under asymmetric excitation results in electromagnetic radiation.

In the next phase of test, additional asymmetry was introduced into the SAW filter, where both the input terminals, i.e. the input electrode and the input ground, were shorted together, as illustrated in figure 6.7. This resulted in an enhanced efficiency of up to 60%, which is shown in figure 6.8. Shorting of the electrodes resulted in a better impedance matching between the source of input excitation and the SAW filter.

In a related experiment, measurements of radiation efficiency for another SAW filter (dimensions 3 mm × 3 mm × 1 mm) mounted on a ground plane made of FR4 (10 cm × 3 cm) with a good return loss in the range of 1800–1885 MHz were taken. The results are illustrated in figure 6.9. In order to further validate the results, further tests were performed on custom-made SAW devices fabricated by Euroquartz. The device was mounted on a board of dimensions 7 cm × 3 cm as shown in figure 6.10. It was comprised of a piezoelectric thin film with interdigital electrodes developed over a two-dimensional area of 3 mm × 3 mm. Its total dimensions while considering the exciting leads was 9 mm × 7 mm × 1 mm. It was designed to have a finite return loss in the range of 1575 MHz. The radiation efficiency during asymmetric excitation while using matched elements was found to be above 60%. When the SAW filter was

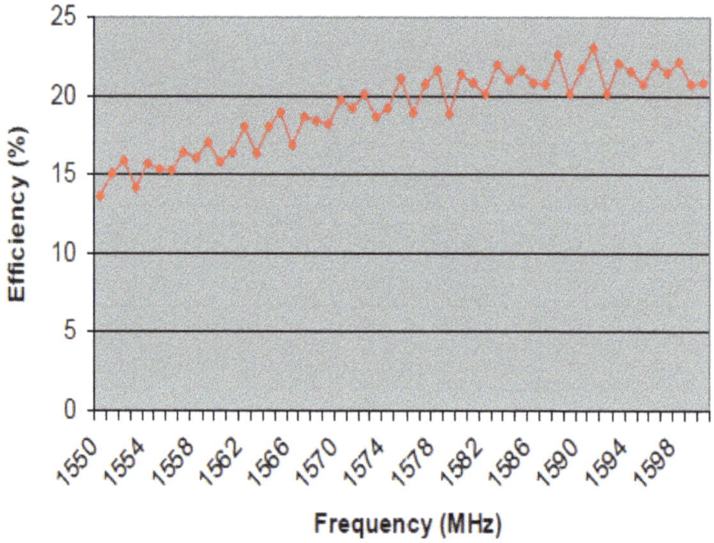

Figure 6.6. The radiation efficiency dropped by 20% as the SAW devices were removed from the antenna board with the track of wire, along with related impedance matching elements on the board at GPS frequencies. Reprinted with permission from [3], copyright (2015) by the American Physical Society.

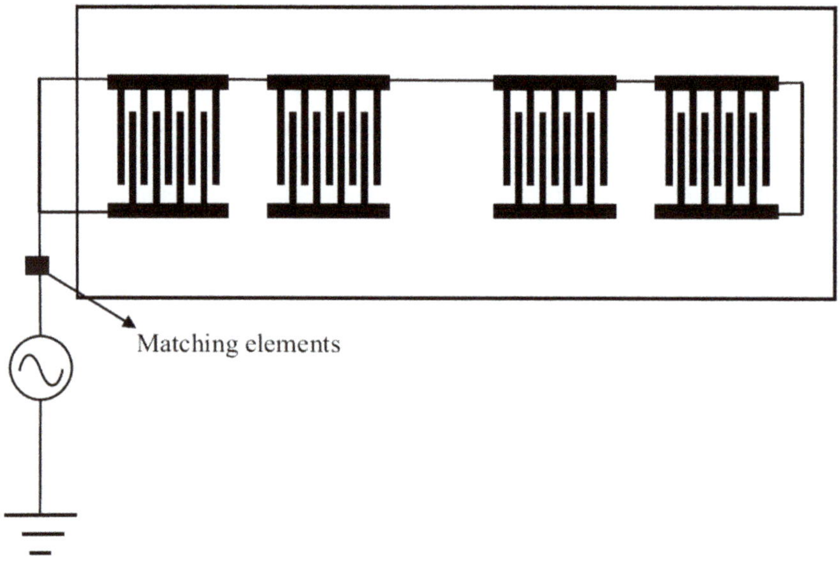

Figure 6.7. The SAW filters as illustrated in figure 6.6 mounted on a board with input terminals shorted in order to illustrate the impact of the excitation pattern on radiation efficiency.

removed (while leaving the bare exciting lead of length 0.9 cm with suitable impedance matching elements), the efficiency dropped to 40%. The red curve in figure 6.11 illustrates the drop in efficiency, while further validating the fact that the SAW device plays a key role in radiation under asymmetric excitation.

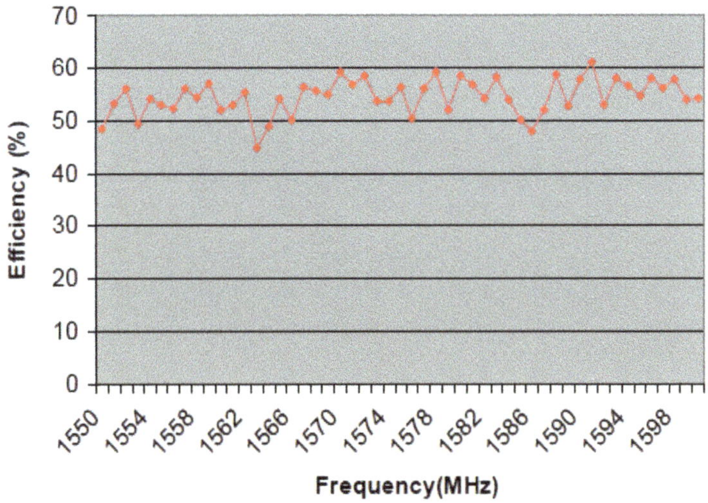

Figure 6.8. There is an increase in radiation efficiency in the range of 60% as the input terminals of the SAW filters are shorted. Reprinted with permission from [3], copyright (2015) by the American Physical Society.

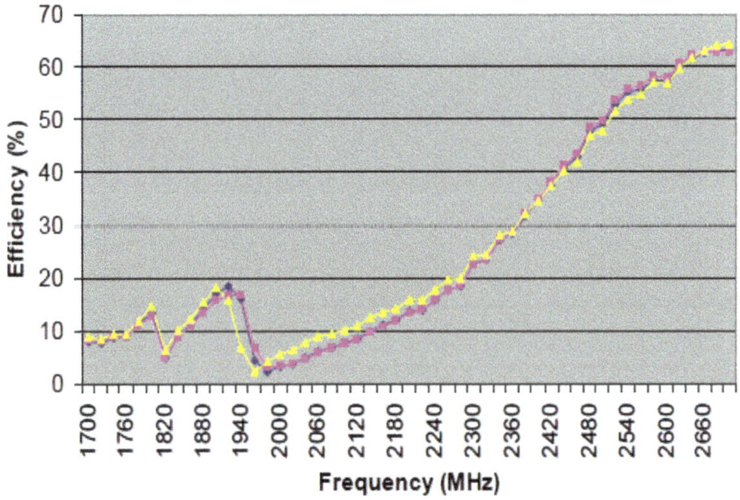

Figure 6.9. Radiation efficiency measurements for a SAW filter operating at 1800–1885 MHz without matched elements. Reprinted with permission from [3], copyright (2015) by the American Physical Society.

The radiation efficiency of SAW filters drops significantly when the electrodes are subjected to symmetric excitations where the input electrode is connected to the voltage source and the ground terminal is effectively grounded. It is ensured that no terminal is left floating in free space. Under such an excitation mechanism, the SAW filters do not show antenna-like features, but there is still some radiation in the near field region [4].

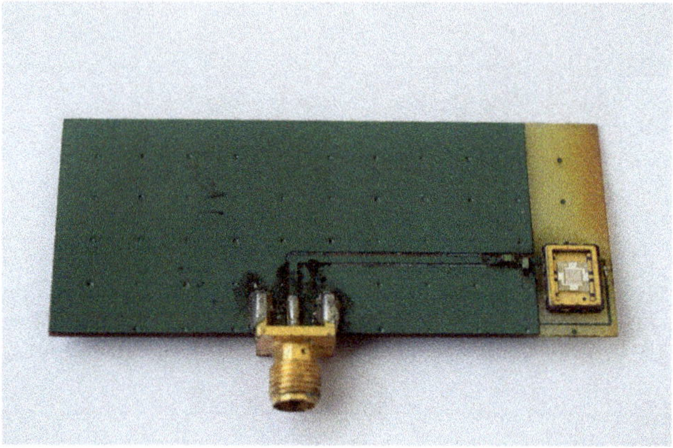

Figure 6.10. A SAW antenna mounted on an antenna board of dimensions 7 cm × 4 cm.

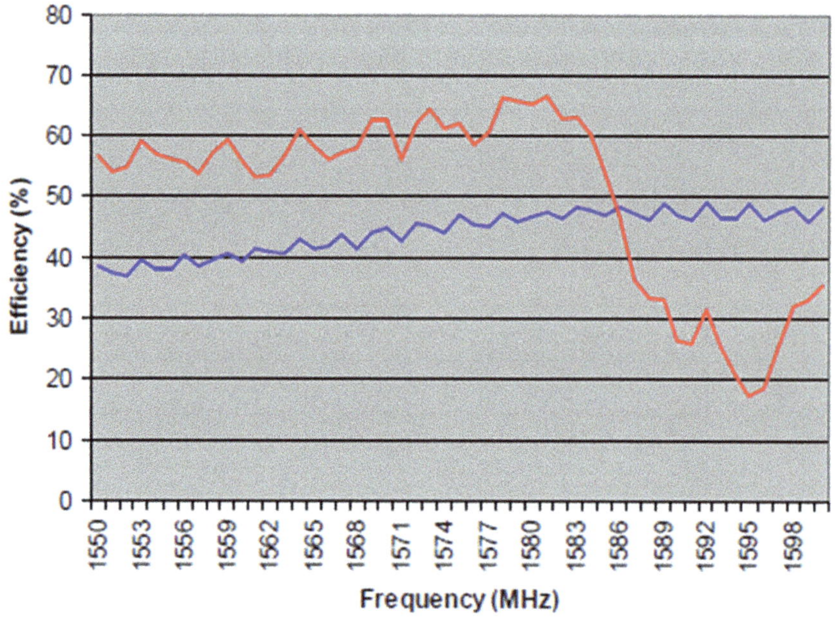

Figure 6.11. Radiation efficiency measurements with a custom-made SAW filter mounted on a board (red line). The drop in radiation efficiency at 1575 MHz as the SAW device was removed indicates its role in radiation. The blue line shows the measurements without the SAW filter. Reprinted with permission from [3], copyright (2015) by the American Physical Society.

The experiments demonstrate that asymmetric excitation fed to thin films of piezoelectric materials at the UHF range results in high radiation efficiency, which is comparable to other materials with high permittivity and comparable physical dimensions. An advantage associated with piezoelectric materials is the fact that they can be fabricated in thin film forms of micrometric thicknesses using conventional fabrication technologies.

In a related experiment, electromagnetic field coupling between a stack of piezoelectric material and the magnetic field associated with an RF coil excited at the medium frequency range of the radio spectrum was measured in the near field region [5, 6]. The RF magnetic field induced a voltage across the leads of the piezoelectric stack that resulted in mechanical vibrations in the material. A microcantilever mounted on the piezoelectric stack amplified and filtered the vibrations, which were measured using a laser–photodetector system. The mechanism of the voltage induction in a dielectric material offered initial insights, which were later leveraged in the context of thin film-based piezoelectric materials.

The theoretical arguments and empirical observations lead us to the conclusion that when the symmetry of an electrodynamic system is broken in an explicit manner, it is reflected in the non-invariance of the associated Lagrangian along temporal as well as spatial dimensions. Under time-varying electrical excitation, such a system with broken symmetries emits electromagnetic radiation due to the explicit symmetry breaking of the electric field and non-conserved Noether current in the localised region of space and time.

This work proves that explicit symmetry breaking of an electrodynamic system and the associated symmetry breaking of the electric field results in electromagnetic radiation. There is a non-conservative Noether current within a localised region of the corresponding system which is transformed into electromagnetic radiation. The ideas are described in detail while using the example of a parallel two-wire transmission line excited by a voltage source at one end which radiates electromagnetic waves only when the lines are flared towards the end and the translational symmetry of the electric field is explicitly broken and overall current in the wire is not conserved. The mechanism of radiation from an asymmetrically excited capacitor–inductor filter is also illustrated in order to justify the theoretical angle. The concept of radiation under explicit symmetry breaking is also used to explain the phenomenon of radiation from DRAs while highlighting the fact that the current literature has some fundamental theoretical gaps. Empirical results on radiation efficiency measurements from asymmetrically excited piezoelectric filters are presented in order to explore an alternative but closely related system. The ideas should open up novel avenues for antenna design, leading to their prospective miniaturisation and eventual integration on an electronic chip.

References

[1] Smith W R, Gerard H M, Collins J H, Reeder T M and Shaw H J 1969 *IEEE Trans. Microw. Theory Tech.* **17** 865

[2] Campbell C K 1998 *Surface Acoustic Wave Devices for Mobile and Wireless Communications* (Amsterdam: Elsevier)

[3] Sinha D and Amaratunga G A J 2015 *Phys. Rev. Lett.* **114** 147701

[4] Sinha D 2009 *US Patent Specification* 12/809,780; *WIPO Patent No. 2009081089*

[5] Sinha D, Pisana S and Flewitt A J 2011 *Smart Mater. Struct.* **20** 025016

[6] Sinha D 2009 *PhD Thesis* University of Cambridge

IOP Publishing

Explicit Symmetry Breaking in Electrodynamic Systems and Electromagnetic Radiation (Second Edition)

Dhiraj Sinha and Gehan A J Amaratunga

Chapter 7

Radiation from a superconducting loop

7.1 Superconducting antennas

For some specialized applications, superconductors can be used as transmitting and receiving antennas considering the fact that they offer size reduction due to high current density and zero dissipation [1, 2]. Superconducting transmitters store energy in the circulating currents [3], which enables them to meet the requirements of peak transmitting power at high efficiency [4].

In a recent work carried out at the University of Cambridge by John Blumer *et al* [5], controlled GHZ wide band radiation was obtained by integrating an antenna, a waveguide, SMEs and a photo-switch into an integrated transmitter device. The main component of the device comprised a superconducting ring made of thin films with size of the order of a couple of centimeters in diameter. Current was generated in the ring and it was excited by a pulse of laser light from an external source. As a result of it, electromagnetic radiation was emitted in the GHz range which was measured in the far-field domain. The radiation spectrum was found to be governed by the dimensions of the superconducting ring, and the nature of laser excitation and position of illumination along the ring. The radio emission spectrum could be tuned by modulating the pattern of laser illumination along the superconducting ring.

This technique illustrates some of the initial results on generation of radiation from high speed transient currents in a direct current (dc) circuit. Conventional antenna theory does not address the issue of radio signal generation from transients existing in dc circuits. This could be explained from the perspective of electromagnetic radiation under explicit symmetry breaking of the electric field [6]. Under laser excitation, the symmetry of the electric field associated with Cooper pairs is explicitly broken during the formation of quasi-particles and their recombination into the original Cooper pairs. As the device was fully detached from metallic wires or related conductive components, the possibility of radiation from secondary effects was assumed to be negligible.

doi:10.1088/978-0-7503-5132-4ch7

7.2 Experimental setup

In the experiments carried out by Bulmer *et al*, the outer surface of a superconducting ring made of intrinsically doped yttrium barium copper oxide (YBCO) was mounted on sapphire substrate (figure 7.1). It was illuminated with a laser pulse lasting 120 fs. The laser was supplied by Specra Physics Hurricane Ti: Sapphire which provided a wavelength of 808 nm. The typical value of energy per pulse was 1 mJ over 120 fs time span. According to prior work, it is known that laser pulses of femtosecond duration result in development of voltage transients lasting over a period of pico-seconds in the superconducting ring as a consequence of excitation of Cooper pairs into quasi-particles and their eventual recombination [7–9].

The YBCO rings were manufactured by Star Cryoelectronics using conventional microfabrication processes. The thickness of YBCO film was 330 nm on an *R* plane sapphire wafer coated with CeO_2 having a thickness of 40 nm (figure 7.1(a)). The critical current was 3.4 $MAcm^{-2}$ and the critical temperature was 87.8 K. The ring was pasted on a sapphire rod with a cryogenic epoxy which was thermally conductive. The components were placed in a Styrofoam container which had liquid nitrogen at 77 K in order to provide cooling. The ring was kept moisture-free with nitrogen boil.

Current was set up in superconductors by cooling the material below its critical temperature in the vicinity of a magnetic field. As the material became a superconductor below its critical temperature, the magnet was removed, which induced current in the superconducting ring. The technique has some disadvantages with regard to speed. A faster way to induce current in the given system is by firing a pulse of laser light on the ring immersed in an external magnetic field generated through a solenoid. The solenoid had a length of 6.75 cm and was wound into 48 turns. Under

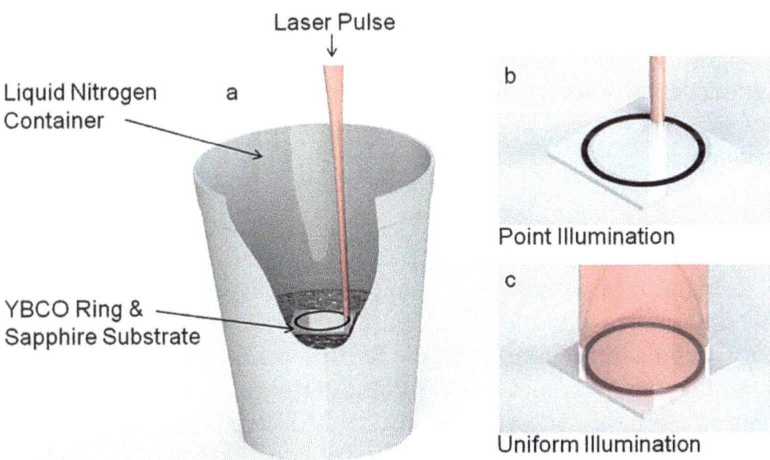

Figure 7.1. (a) Experimental setup illustrating a superconducting ring made of YBCO on a sapphire substrate illuminated by laser light. (b) The superconducting ring was illuminated at a point, (c). The superconducting ring was illuminated uniformly around the ring. Reprinted from [5], copyright (2015) with permission of Springer.

the presence of a magnetic field, the superconducting ring develops a screening current in order to oppose any change in magnetic flux through its centre. The ring's screening current was interrupted by the laser pulse excitation. This allowed penetration of magnetic flux into the central portion of the superconducting ring. The magnetic flux was found to remain trapped in the ring even after the solenoid, which was the source of magnetic field in the experiment, was removed, leading to generation of superconducting current in the ring which continued to flow without any dissipation. The value of the current was calculated by measuring the magnetic field in the ring using Hall probe and using Biot–Savart law.

When the superconducting ring with a finite value of current was illuminated by another laser pulse, it resulted in generation of electromagnetic radiation. The resulting signal was detected using far field antennas. Two different antennas were used. The first antenna was a horn antenna positioned in the plane of the ring at a distance of 1.6 m, which was the far field domain for the given frequency ranges. It was calibrated to operate between 1–18 GHz. Another antenna (a biconic/log-periodic hybrid antenna with calibrated gain curve 25 MHz–7 GHz) was placed at an angle of 45 degrees from the ring's plane at a distance of 1.9 m in order to monitor out-of-plane radiation. The antennas were connected to digital oscilloscopes with calibrated cables (Lecroy 65 GHz LabMaster and Tektronix 33 GHz DPO73304DX). The amplitude of voltage measured by these instruments was used to calculated the energy spectra by using the Friis equation while including the gain and loss associated with the antenna and the feeding cables.

In order to isolate the radiating system from other electromagnetic field effects, metallic objects were removed around the ring within a radius of 45 cm. This ensured that the near-field region of the superconducting antenna was free from secondary sources of radiation. However, some tests were also carried out with metallic objects near the ring in order to gauge their exact role in influencing the radiation characteristics.

The experiments were not connected in an anechoic chamber, and so the electromagnetic environment of the laboratory was measured by recording the baseline signals. Microwave signals are also produced by lasers and other electro-optic components.

7.3 Results

The time domain representation of the electromagnetic signal emitted by the superconducting ring is illustrated in figure 7.2. The signal lasts over a period of 2 ns, which is significantly larger than the femtosecond level of laser excitation. The experimental setup was conducted on an optics table having a metallic surface which was 45 cm away from the superconducting ring. Electromagnetic waves need around 3 ns to traverse the round-trip distance between the ring and the metallic surface. This precludes the possibility of generation of electromagnetic waves by a secondary effect caused by the metallic part of the table in the 3 ns period. However, the signal components in the spectrum after 3 ns might include multipath reflections. A problem which was encountered during the experiment was linked to laser induced

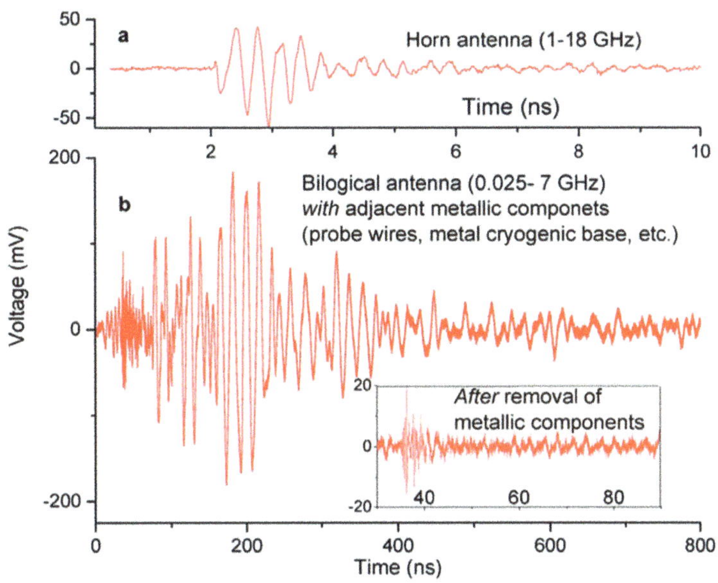

Figure 7.2. Microwave generation by a superconducting ring recorded by an oscilloscope. (a) Signal recorded by a horn antenna (1–18 GHz) oriented along the plane of superconducting ring with isolation from metallic components. (b) Signal recorded by a Bilogical antenna (0.025–7 GHz) without isolation of the ring from metallic components. Inset shows the signal picked up by the same antenna with ring isolated from metallic components. Reprinted from [5], copyright (2015) with permission of Springer.

deterioration of YBCO while it was focused on a particular point along the ring with a layer of liquid nitrogen in order to maintain low temperature. However, in most cases, the YBCO ring was found to support super-current long after the pulses were switched off.

Before the laser excitation, the magnetic field in the ring was measured in order to confirm the presence of superconducting current. After the microwave measurements, it was found that the magnetic field remained in the range of 2–10% of its original strength.

During the initial phase of the experiments, an inductive coil was left above the ring in order to energize it. Besides this, the temperature sensing and field measuring probes were fixed on the sapphire substrate. The signal received under such a condition is illustrated in figure 7.2(b). In a related experiment, the ring had a discontinuity which was used to feed dc current into it. Such conditions resulted in signals having relatively longer wavelengths, and durations larger by a factor of two in comparison to that of the isolated case. It appears that the metallic probes increased the overall size of the antenna and increased the intensity of multi-path reflections. However, in the second phase of the experiment, the ring was isolated from all the metallic components.

In the next phase of the experiments, the radiated spectrum was measured by increasing the amount of current in the ring. The results are illustrated in figure 7.3(a)–(c). The spectrum of radiation under an increasing value of current is

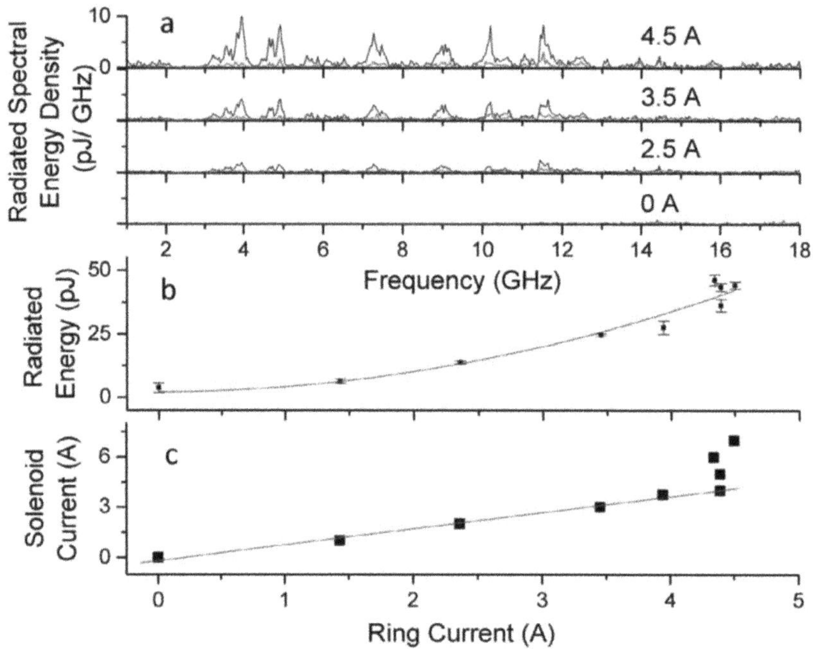

Figure. 7.3. Dependence of radiation on current in superconducting ring. (a) Spectral energy density with a change in current. The red curve indicates standard deviation over a number of measurements. (b) Change in energy of microwave with an increase in superconducting current in ring. (c) Linear change in ring current with an increase in solenoid current. Reprinted from [5], copyright (2015) with permission of Springer.

shown in figure 7.3(a). There is a linear relationship between energy of the radiated signal and magnetic energy of the ring which is proportional to the square of current. This is confirmed by a quadratic increase in the value of energy with current, as illustrated in figure 7.3(b). The proportionality relationship between the current in the solenoid and the ring is shown in figure 7.3(c). The relationship is valid until the critical current in the ring whose value is 4.5 A at 77 K beyond which superconductivity is destroyed.

The magnitude of the signal was measured to be dependent on polarization as well as the energy of the laser pulse, but the spectral form factor was retained. An interesting observation was the fact that a minimum laser intensity threshold was needed in order to generate radiation. However, an increase in the strength of laser intensity by increasing its pulse energy did not increase the amplitude of the radiation. It appears to indicate that the Cooper pairs could only be excited in a specific manner in order to generate radiation.

In the next phase of the experiments, the impact of duration of laser pulse on the emitted radiation was explored. For pulse duration shorter than 500 fs, the output signal was not affected either in terms of its amplitude or frequency profile. When the pulse duration was increased in the range of 1 ps, it reduced the amplitude without influencing the spectral distribution. When the duration of the pulse was 1 ns, no microwave radiation was produced.

7.4 Analysis

Any change in the magnitude of current in the superconducting ring or laser intensity changed the magnitude of the radio signal but it did not influence the spectral distribution which was primarily observed to be governed by the size of the ring. An important observation in the context of laser excitation of superconducting ring was that global illumination did not result in measurable radiation.

The most outstanding observation was that the intensity of microwave signal began to drop as the pulse duration was increased to approximately 1 ps, which is the typical time scale related to pairing and depairing of Cooper pairs. The time scale of laser excitation must be in the range of a couple of hundred femtoseconds in order to allow the excitation of Cooper pairs.

The mechanism of radiation in the given experiments was significantly different from broadband THz radiation from picosecond switching of photo-induced superconducting devices [8, 10, 11]. It could be asserted with certainty that these were not the tail ends of THz radiation as their amplitude did not increase with an increase in laser intensity.

The operation of a superconducting ring as an antenna under laser excitation which excites the Cooper pairs and electromagnetic radiation is a result of recombination of the associated quasi-particles underscores the role of explicit symmetry breaking mechanism of radiation. The radiation is not linked to current modulation associated with a typical RL circuit as the emitted spectrum is closely associated with the physical dimensions of the ring. Only a given set of modes can be excited under laser excitations and this is maintained not only during point illumination but also under uniform illumination of the given structure. Radiation was measured by two different antennas oriented along the plane of the ring and out of plane of the ring. Both the antennas captured a similar radiation spectrum. There was a linear relationship between the energy of the radiated signal and the energy content of the superconducting ring. With a change in current in the ring, the energy of the emitted spectrum changed but there was no change in the spectral distribution which also remained independent of the variation of laser intensities.

References

[1] Khamas S, Cook G, Kingsley S, Woods R and Alford N 1993 Investigation of the enhanced efficiencies of small superconducting loop antenna elements *J. Appl. Phys.* **74** 2914

[2] Lancaster M 1997 *Passive microwave device applications of high temperature superconductors.* (Cambridge: Cambridge University Press)

[3] Ali M *et al* 2010 An overview of SMES applications in power and energy systems *IEEE Trans. Sustainable Energy* **1** 38

[4] Ullrich G 1995 Summary of the DNA SMES development program *IEEE Trans. Appl. Supercon.* **5** 416–21

[5] Bulmer J, Bullard T, Dolasinski B, Murphy J, Sparkes M, Pangovski K, O'Neill W, Powers P and Haugan T 2015 Tunable broadband radiation generated via ultrafast laser illumination of an inductively charged Superconducting ring *Sci. Rep.* **5** 18151

[6] Sinha D and Amaratunga G 2015 Electromagnetic radiation under explicit symmetry breaking *Phys. Rev. Lett.* **114** 147701
[7] Hegman F *et al* 1995 Electro-optic sampling of 1.5-ps photoresponse signal $YBa_2Cu_3O_{7-\delta}$ *Appl. Phys. Lett.* **67** 285
[8] Hangyo M *et al* 1997 THz emission from high-T_c superconductors by optical pulse excitation *IEEE Trans. Appl. Supercon.* **7** 1051–822
[9] Jukna A, Taneda T and Sobolewski R 2004 Control of gigahertz antenna radiation using optically triggered Y–Ba–Cu–O superconducting microbridges *Supercond. Sci. Technol.* **17** S336–40
[10] Tonouchi M and Saijo H 2003 Terahertz radiation from YBCO thin film log-periodic antennas *J. Supercond. Nov. Magn.* **16** 867
[11] Kawayama I, Zhang C, Wang H and Tonouchi M 2013 Study on terahertz emission and optical/terahertz pulse responses with superconductors *Supercond. Sci. Technol.* **26** 093002

IOP Publishing

Explicit Symmetry Breaking in Electrodynamic Systems and Electromagnetic Radiation (Second Edition)

Dhiraj Sinha and Gehan A J Amaratunga

Chapter 8

Broken symmetries in thermodynamics

8.1 Introduction

According to classical thermodynamics, transfer of heat increases entropy of a system which is associated with an increase in the number of microstates or accessible states [1]. The different processes in thermodynamics are assumed to occur under equilibrium conditions which assumes a symmetric energy landscape [2]. However, the internal energy of a set of molecules has kinetic as well as potential energy components and the energy landscape may deviate from a symmetric assumption. The kinetic and potential energy profiles of molecules show different behaviour in terms of geometric distribution in the physical space. A molecule in a relatively higher state of kinetic energy transfers momentum to less energetic molecules leading to distribution of energy, while increasing entropy. But a molecule gaining potential energy is likely to remain confined in the neighbourhood of a central field. For example, a charged molecule can form localized bonds with other molecules while reducing the kinetic energy of a set of other molecules. If such a system, at a given equilibrium is subjected to free expansion, the final equilibrium state would also be defined by the local potential energy gradient, which must be incorporated into a calculation of entropy changes [3].

8.2 Entropy changes under potential gradients

Absolute temperature is a measure of internal energy of a given thermodynamic system [2]. For a given volume, internal energy is directly proportional to the kinetic energy of molecules. Internal energy increases under heat transfer which leads to an increase in temperature in an isochoric process. Thus, entropy changes are directly associated with changes in kinetic energy of molecules.

Transfer of heat energy can also increase the potential energy of molecules which are embedded within a conservative vector field, which can reduce the mobility of

doi:10.1088/978-0-7503-5132-4ch8

molecules in geometric space [4]. A spring mass system of mass m and spring constant k is subjected to a liquid which has highly energetic molecules, some of the molecules would be able to push the mass in a certain direction by exerting a force $\boldsymbol{F}$ and its acceleration can be written as, $md^2\boldsymbol{x}/dt^2 = \boldsymbol{F} - k\boldsymbol{x}$, where $\boldsymbol{x}$ is displacement. The total work can be found by integrating this equation. The integral of force with distance can be linked with the energy transfer from heated molecules δH and the energy stored in the spring is δU (figure 8.1(a)). The total kinetic energy of the molecules is δQ and $\delta Q = \delta H - \delta U$. During an external measurement at a certain instance of time, the energy δQ would appear as the kinetic energy of the system which can be transformed into heat. Molecules having a finite value of potential energy will not contribute to a rise in temperature of the measuring device through energy dissipation, thus, the potential energy term, δU effectively reduces the accessible states of a system if it has a finite gradient along the spatial dimension. Thus, the net change in entropy can be written as [3],

$$\Delta S = \frac{\delta Q}{T} = \frac{\delta H - \delta U}{T} \tag{8.1}$$

The thermal energy excites the mass of the oscillators, while increasing their accessible states, however, at a certain instant of time, the spring elements also store energy while reducing the range of accessible states leading to short-range order. A thermodynamic system is assumed to be in equilibrium and we assume that

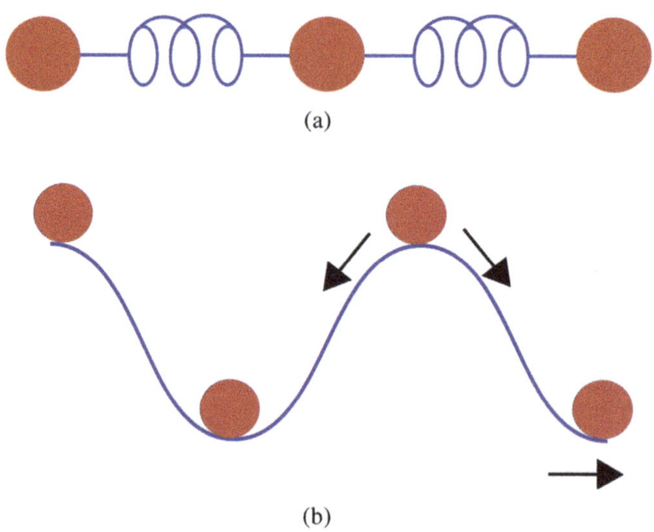

Figure 8.1. Explicit symmetry breaking in potential surfaces. (a) Thermal energy increases the kinetic energy of the mass which is transferred to the spring element. The potential energy gradient of the spring breaks the existing symmetry of the system by offering a constraint on the existing degree of freedom which can reduce entropy. (b) The potential energy profile can act as a barrier towards acquiring accessible states. The accessible states of a ball rolling to the left and the right of an energy landscape would be different due to the differential nature of potential energy landscapes.

the process of induction of potential energy is such that the system has remained in equilibrium during the process of heat transfer.

The potential energy and associated conservative force fields can act as a barrier in terms of constraining the degrees of freedom of gas molecules. Figure 8.1(b) illustrates a related case, where the motion of a given mass is dependent on the shape of the potential energy landscape within its localized region. This becomes particularly evident in the context of gas molecules where energy is applied in such a manner which causes ionization. For example, if a section of the chamber is negatively charged by inserting an electrode of charge Q, then the gas molecules would be under the effect of electrostatic force fields, which will cause some kind of concentration of charges around the electrodes (figure 8.2). The net potential energy of such a system is,

$$U_e = \sum_{i=1}^{n}\sum_{j=1}^{n}\frac{q_j q_j}{r_{ij}} - \sum_{i=1}^{n}\frac{q_i Q}{r_i} \tag{8.2}$$

where r_{ij} is the distance between the ith charge q_i and jth charge q_j and r_i is the distance between charge q_i and the electrode having a net charge Q. Here, the molecules behave like harmonic oscillators under the influence of electrodynamic

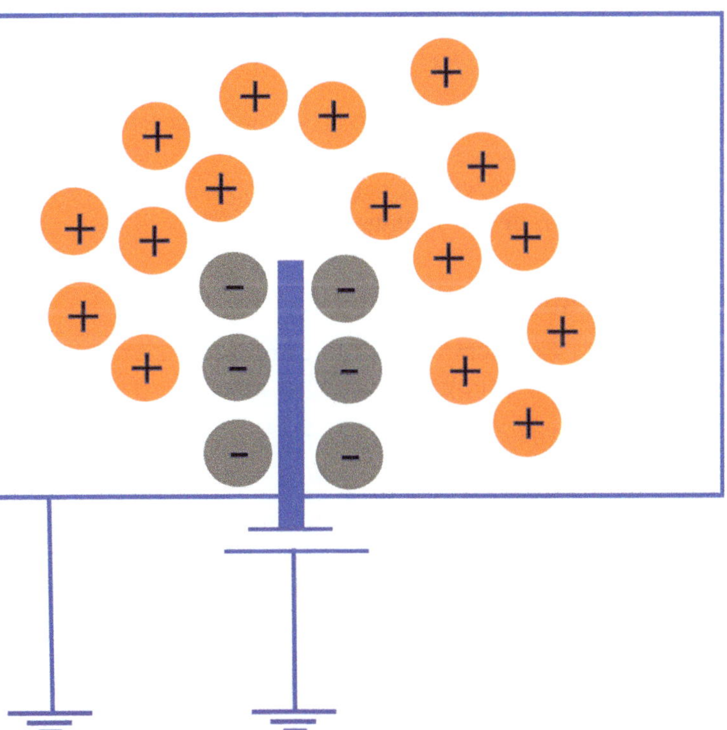

Figure 8.2. The role of potential gradients in entropy reduction The electric potential of the electrode reduces the degree of freedom of molecules and reduces entropy.

force fields. The presence of conservative force fields reduces the overall entropy of the system.

The concept is general and applies to other related interactions, for example, when photons excite a system, there is generation of heat as the photons are absorbed by the system. However, a part of the input energy of photons is spent in meeting the threshold energy of the electrons, and the overall change in entropy is expressed by [3],

$$\Delta S = \frac{h\nu - h\nu_0}{T} \tag{8.3}$$

where h is the Planck's constant, ν is frequency of photon and ν_0 is the threshold frequency of emission. The entropy of a thermodynamic system with a total number of accessible states $\Omega(\mathcal{E})$, which is dependent on energy $\mathcal{E}$ is expressed by, $S = k_B \ln \Omega(\mathcal{E})$ where K_B is Boltzmann constant [1]. Under the effect of heat transfer there is a change in the number of accessible states and the net entropy change is obtained using Taylor expansion,

$$\Delta S = k_b \ln \Omega(\mathcal{E} + \Delta H - \Delta U) - k_B \ln \Omega(\mathcal{E}) = k_B\left(\frac{\partial \ln \Omega}{\partial \mathcal{E}}\right)(\Delta H - \Delta U) + K\frac{1}{2}\frac{\partial^2 \ln \Omega}{\partial \mathcal{E}^2}(\Delta H - \Delta U)^2 + \cdots \tag{8.4}$$

Substituting $\partial \ln \Omega(\mathcal{E})/\partial E$ by η, which is a constant, we get,

$$\Delta S = k_B \ln \Omega(\mathcal{E} + \Delta H - \Delta U) - k_B \ln \Omega(\mathcal{E}) = k_B\eta(\Delta H - \Delta U) + k_B\frac{1}{2}\frac{\partial \eta}{\partial \mathcal{E}}(\Delta H - \Delta U)^2 + \cdots \tag{8.5}$$

The thermal reservoir is assumed to be much larger in comparison to the thermodynamic system, which implies that η remains invariant. Thus, the higher order terms in the expansion can be neglected and we after substituting η by $1/(k_B T)$, the prior expression for entropy change is obtained.

8.3 Entropy production

Systems driven far away from equilibrium produce entropy at a given rate [5]. At equilibrium the rate of entropy production is zero,

$$\frac{dS}{dt} = \frac{1}{T}\frac{d(H - U)}{dt} = 0 \tag{8.6}$$

The above relationship implies that a finite rate of increase of potential can reduce the net rate at which entropy is being produced from a system. In case the rate of changes in potential is equal to the thermal energy input, the system, as a thermodynamic unit would appear to be in equilibrium even under steady heat transfer.

At a given equilibrium the entropy change is non-zero across the spatial coordinates. When heat is transferred under non-equilibrium conditions, there is a sudden rise in potential as well as kinetic energy gradient. The entropy gradient along a given direction r can be written as,

$$\frac{dS}{dr} = \frac{1}{T}\frac{d(H - U)}{dr} \tag{8.7}$$

It implies that the potential gradients can also play a key role in reduction of entropy gradient [3]. Under specific conditions, the gradient of potential and kinetic energy could be balanced leading to a net zero value of entropy gradient, and the system could appear to be in equilibrium.

An isolated thermodynamic system evolves towards a state of maximum entropy. In fact, it has the tendency of evolving towards a maxim level of symmetry. At equilibrium, there is maximum entropy as well as symmetry. This is evident from the fact that the molecules pass through a set of microstates having specific energy levels which are equiprobable. Equation (8.7) indicates that the state of maximum entropy is realized only when the potential energy gradient is zero. This is also supported form the principle of classical thermodynamics, which assumes absence of force fields.

8.4 Entropy changes under symmetry breaking

In the context of a crystalline material, heat removal results in phase transition at a critical temperature where order is generated associated with the process of spontaneous symmetry breaking [6]. Classical thermodynamics does not address entropy changes in such systems which are driven away from equilibrium. Some novel physical and mathematical framework is needed in order to address entropy changes in such systems, which have been recently reported [7]. It offers some novel theoretical trajectories towards understanding the phenomenon of studying entropy changes in non-equilibrium systems.

The calculations on entropy changes in non-equilibrium systems are focused on entropy production under dissipation [5]. Phase transition in a crystalline material is a non-dissipative process which needs a novel physical and mathematical framework. It should incorporate the nature of changes in electromagnetic potentials in a crystal and their inherent symmetries which undergo a change [7]. In a crystalline material like a ferroelectric or ferromagnetic material, there is no net magnetism at room temperature. We can say that the magnetic or electric dipole moments in such systems have rotational symmetry. Below a critical temperature, the energy landscape goes through spontaneous symmetry breaking (figure 8.3(a)) and an order parameter is generated, which is electric dipole moment in a ferroelectric material and a magnetic dipole moment in a ferromagnetic material (figure 8.3(b)) [8].

The energy landscape under broken symmetry is given by [8],

$$V = \beta^2\phi^*\phi + \frac{1}{2}\lambda(\phi^*\phi)^2 \tag{8.8}$$

where $\phi = (\phi_x + i\phi_y)/\sqrt{2}$ and $\beta = a(T - T_c)$, a is a constant, T_c is critical temperature and λ denotes a dimensionless quantity with a positive value. The term, $\lambda(\phi^*\phi)^2/2$ indicates that the potential function has global phase symmetry. For, $\lambda < 0$, $\phi = \phi_0$, and,

$$\phi_0^*\phi_0 = \beta^2/\lambda \tag{8.9}$$

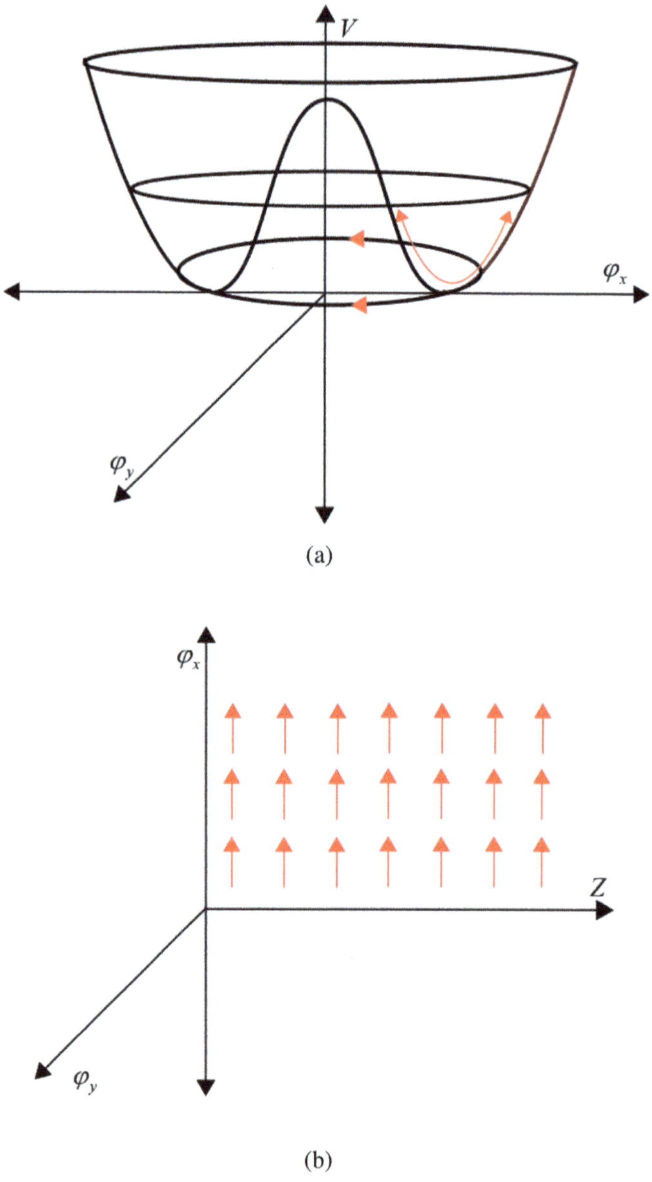

Figure 8.3. Potential energy profile during symmetry breaking. (a) The potential energy profile is induced at critical temperature below which a net magnetization is induced and the symmetry is spontaneously broken. There are modes along the axial and radial directions. (b) The rotational symmetry of the magnetic dipole moments is lost as the symmetry is broken. Reprinted from [7], copyright (2022), with permission from Elsevier.

It denotes a circle in the ϕ plane, with a radius r and $\phi_0 = \beta e^{i\theta}/\sqrt{\lambda} = re^{i\theta}$. It corresponds to a degenerate state having a number of ground states corresponding to different values of θ. The system has made a transition from its initial state of symmetry to a final state of broken symmetry. Particles in such a low state can propagate with low energy excitations and they are called Goldstone modes (figure 8.4(a)).

Calculations on entropy changes in such systems must include the changes in internal energy as a consequence of heat transfer, which can excite the amplitude of the order parameter, which could be magnetic waves in ferromagnetic materials or

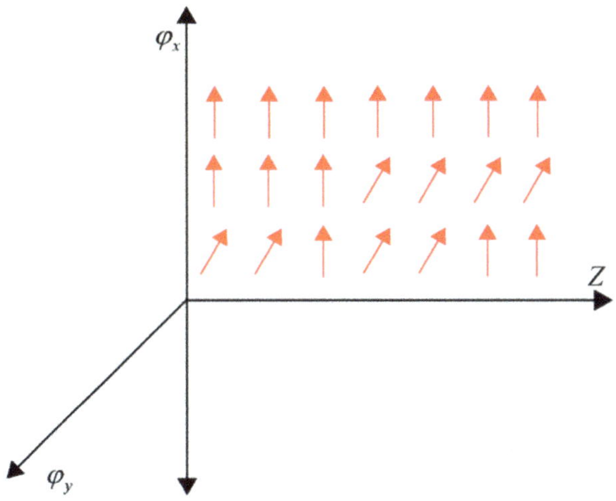

(a)

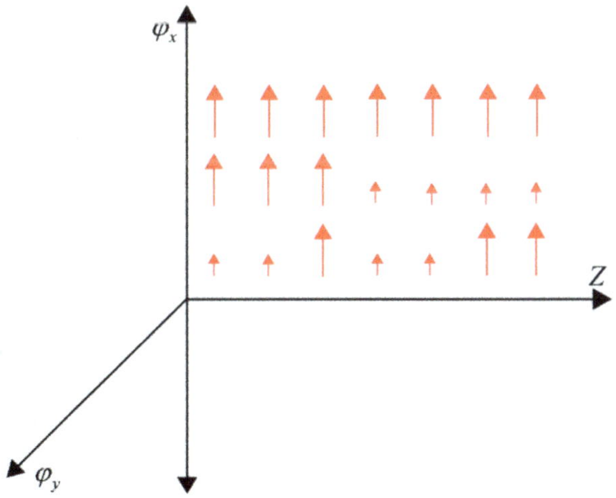

(b)

Figure 8.4. Goldstone- and Higgs-like modes. (a) Excitations in the ϕ_x direction denote Higgs-like modes which are massive. (b) Modes propagating along the ϕ_y direction are low energy modes called Goldstone modes. Reprinted from [7], copyright (2022), with permission from Elsevier.

polarization waves in ferroelectric materials. These modes have the capability of propagation along radial as well as axial directions of the potential energy surface. At a mathematical level, they correspond to $\phi = \phi_0 + \chi$, where $\phi_0 = \beta/\sqrt{\lambda}$ and χ represents a dynamical field. The potential energy function is,

$$U = U(\phi_0) + \frac{1}{2}\lambda\left[\phi_0(\chi^* + \chi) + (\chi^*\chi)\right]^2 \tag{8.10}$$

Using $\lambda\phi_0^2 = \beta^2$ and, $\chi = (\chi_1 + i\chi_2)/\sqrt{(2)}$ leads to

$$U = U(\phi_0) + \beta^2\chi_1^2 + O(\chi^3) \tag{8.11}$$

where, $O(\chi^3)$ denotes the higher order terms. With $\phi = 0$, the system lies at a vacuum state with a symmetry along the radial directions. For a specific value of $\phi = \phi_0$, the symmetry is explicitly broken and $\beta/\sqrt{\lambda}$ varies. Under such a condition, for the field, χ_1, massive modes having angular frequency, $\omega = \sqrt{(k^2 + 2\beta^2)}$, with k as the wave number are created. It changes the magnitude of magnetic dipole moments as illustrated in figure 8.4(b). These modes behave as if they have a finite mass. The entire field oscillates like a homogenous entity with an infinite wavelength.

Thus, transfer of heat can generate fluctuations in the magnetization associated with massless as well as massive modes which can offer new insights in calculations on entropy changes. The generation of Higgs-like modes can annihilate a part of the thermal energy. These modes can also annihilate the Goldstone modes. The net effect of it is a reduction in entropy as an energy gap ΔU is developed. Thus, the net entropy change can be expressed as, $\Delta S = \int \delta Q/T = (\Delta H - \Delta U)/T$ [7].

8.5 Entropy reduction in dissipative systems

According to Prigogine, a dissipative system is driven away from equilibrium and produces entropy [5]. He underscored the importance of fluctuations in entropy production and the onset of irreversibility in non-equilibrium systems. He further noted that such systems are also associated with self-organization and order. It remains a mystery how dissipative systems can generate order while continuing to produce entropy. This question can be answered if we consider the potential energy changes and conservative force fields induced in such systems under non-equilibrium conditions. Prigogine has argued that perturbation of a system around its equilibrium leads to changes in its stability, which can be described using a Lyapunov function [5]. We can always link a Lyapunov function with a potential energy function or an energy gap $U(r)$. We have discussed how massive modes can be formed in such systems under energy exchange, which can reduce entropy and induce order. The snapshot of a related mode of oscillation of magnetic moment in a thin film of ferromagnetic material is shown in figure 8.5 under a high value of damping ratio, which can generate a specific pattern. The potential energy component of Goldstone modes can also excite other modes leading to the formation of more complex patterns.

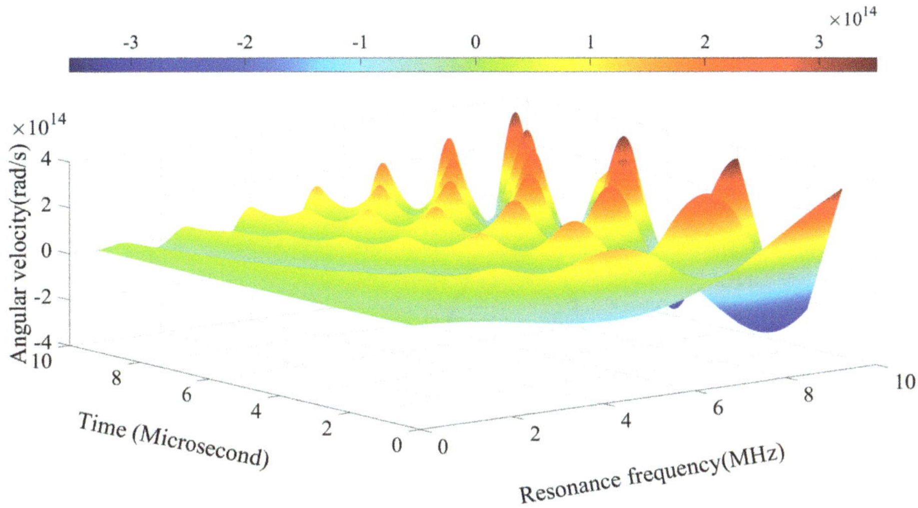

Figure 8.5. The figure illustrates a Goldstone mode which generates a specific pattern at a given instant of time. Here, the evolution of the wave over a period of 0–10 ms is shown with its resonant modes lying between 0.6 MHz and 10 MHz. Each of the resonant modes generate a crest in the pattern. Reprinted from [7], copyright (2022), with permission from Elsevier.

Stochastic formulations have been applied in recent years to describe entropy changes at microscopic level, where fluctuations tend to reduce entropy over short time and length scale. In that context, the fluctuation theorem argues that in a non-equilibrium system, the probability of entropy production is $P(S = A)/P(S = -A) = e^{At}$, where $S = A$ is entropy production with a positive rate A and $S = -A$ is probability of negative entropy production [9]. It implies that entropy can get reduced over short time scales, but it will increase under steady state conditions. The ideas have been proved empirically in the context of trapped beads in fluid excited by optical tweezers which show entropy reduction over short time scales [10]. An optical tweezer essentially generates a force and a localized energy gradient depending on the intensity of optical signal, which can generate a finite amount of work in reducing entropy over short time scales (figure 8.6) [11].

When an optical beam levitates a particle in a fluid, there is energy exchange between conservative force fields with dissipative elements. An additional problem is that the distinction between work and heat can fade in such systems. Hence, interpretation of entropy reduction under fluctuations can also be considered from the perspective of work generation driven by the potential energy gradients. The physical scenario could also be considered from the perspective of generation of Goldstone- as well as Higgs-like modes, although more research would be needed to explore it. Stochastic resonance is another phenomenon where there is generation of order under non-equilibrium conditions [12]. It occurs in bistable systems where the particle can make a transition from one energy state to the other using a potential barrier. A related example is ring laser, where there is signal enhancement when two laser beams propagating in opposite directions are modulated by an acousto-optical

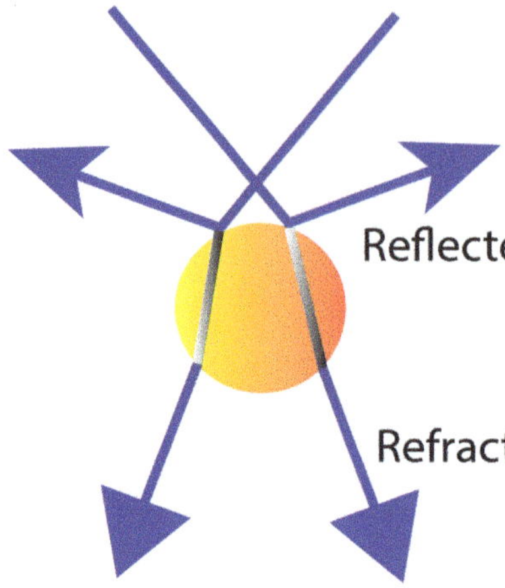

Figure 8.6. A spherical bead made of dielectric material in an optical beam. The light rays impart a net momentum to the bead under refraction and reflection [11].

modulator. In fact the potential energy element of the modulator, offers an energy gradient, which can absorb noise while enhancing the signal.

8.6 Conclusion

Heat energy applied to a system increases its internal energy which changes the kinetic as well as potential energy of the molecules. Thus, the net energy can be defined in terms of its Hamiltonian. Classical thermodynamics is equilibrium thermodynamics which neglects the role of conservative force fields within the system. It assumes the condition of a symmetric energy landscape devoid of potential gradients. However, in reality, potential gradients exist, which can break the symmetry of the system leading to a net reduction in the number of accessible states and its overall entropy [3]. Thus, entropy changes can be expressed in terms of differential changes in Hamiltonian and Lagrangian of the system at a given temperature. It implies that entropy production rate can be slowed down as a consequence of symmetry breaking.

Heat energy applied to a material under the condition of broken symmetry can excite Goldstone-like modes which increase entropy. Such a process can also excite Higgs-like modes which reduce entropy of the system. Such modes with associated energy gaps can reduce entropy while imparting localized order [7]. It offers a novel line of thinking towards understanding entropy reduction in optical tweezers and in a phenomenon like stochastic resonance.

References

[1] Clausius R 1850 *Poggendorff's Ann. Phys., Lpz* **155** 368
[2] Reif F 1965 *Statistical and Thermal Physics* (New York: McGraw-Hill)
[3] Sinha D 2014 Entropy changes in a thermodynamic process under potential gradients *Physica* **416** 676–83
[4] Goldstein H 2002 *Classical Mechanics* (Reading, MA: Addison-Wesley)

[5] Prigogine I 1961 *Non Equilibrium Statistical Mechanics* (New York: Interscience)
[6] Fröhlich J, Simon B and Spencer T 1976 Phase transitions and continuous symmetry breaking *Phys. Rev. Lett.* **36** 804
[7] Sinha D and Bouffanais R 2022 Entropy changes in crystalline material under phase transition and symmetry breaking *Physica* **588** 126525
[8] Landau L D 1969 *Selected Works* vols 1 and 2 (Moscow: Nauka) (in Russian); Collected Papers of L D Landau, edited by D ter Haar (London: Pergamon, 1965)
[9] Evans D J, Cohen E G D and Morriss G P 1993 *Phys. Rev. Lett.* **71** 2401
[10] Wang G M, Sevick E M, Mittag E, Searles D J and Evans D J 2002 *Phys. Rev. Lett.* **89** 050601
[11] Bustamante C J, Chemla Y R, Liu S and Wang M D 2021 Optical tweezers in single-molecule biophysics *Nat. Rev. Methods Primers* **1** 1–29
[12] McNamara B and Wiesenfeld K 1989 Theory of stochastic resonance *Phys. Rev.* A **39** 4854

www.ingramcontent.com/pod-product-compliance
Ingram Content Group UK Ltd.
Pitfield, Milton Keynes, MK11 3LW, UK
UKHW051848100926
8556IPUK00005B/520

9 780750 351300